Using Postmodern and Poststructural Approaches in Applied Research

Using Postmodern and Poststructural Approaches in Applied Research introduces the complex theoretical approaches of postmodern and poststructural thought in ways that are accessible and understandable.

The book begins by grounding our discussion with the foundational work of scholars who first wrote about postmodern and poststructural approaches. We then follow with examples of how scholars continue to use these approaches and theories today and apply them in different practice areas. Examples are provided from the authors' own research, teaching, and mentorship along with activities for the reader, to support them with their own research and application of postmodern and poststructural thought. Interviews and reflections are shared from students and researchers who have used postmodern and poststructural approaches in their own work.

This book is for students and researchers who want to use postmodern and poststructural approaches to make a difference in practice areas where they work or have connections. It intends to inspire, excite, and support students, academics, researchers, and practitioners to use postmodern and poststructural concepts in their everyday practices.

Julianne Cheek is a professor at Østfold University College, Norway.

Megan Aston is a professor at Dalhousie University, Nova Scotia, Canada.

Developing Qualitative Inquiry

Series Editor: Janice Morse
University of Utah

Books in the *Developing Qualitative Inquiry* series, written by leaders in qualitative inquiry, address important topics in qualitative methods. Targeted to a broad multi-disciplinary readership, the books are intended for mid-level to advanced researchers and advanced students. The series forwards the field of qualitative inquiry by describing new methods or developing particular aspects of established methods.

Other volumes in this series include:

Playbuilding as Qualitative Research
A Participatory Arts-Based Approach
Joe Norris

Poetry as Method
Reporting Research Through Verse
Sandra L. Faulkner

Mixed Method Design
Principles and Procedures
Janice M. Morse

Autoethnography as Method
Heewon Chang

Using Postmodern and Poststructural Approaches in Applied Research
Connecting Theory, Method, and Practice
Julianne Cheek and Megan Aston

For a full list of titles in this series, please visit www.routledge.com/Developing-Qualitative-Inquiry/book-series/DQI

Using Postmodern and Poststructural Approaches in Applied Research

Connecting Theory, Method, and Practice

Julianne Cheek and Megan Aston

Routledge
Taylor & Francis Group

NEW YORK AND LONDON

First published 2024
by Routledge
605 Third Avenue, New York, NY 10158

and by Routledge
4 Park Square, Milton Park, Abingdon, Oxon, OX14 4RN

Routledge is an imprint of the Taylor & Francis Group, an informa business

ISBN: 978-0-367-14883-6 (hbk)
ISBN: 978-0-367-14884-3 (pbk)
ISBN: 978-0-429-05376-4 (ebk)

DOI: 10.4324/9780429053764

Typeset in Times New Roman
by SPi Technologies India Pvt Ltd (Straive)

Contents

Acknowledgements

We would like to thank

- Our mentors who introduced us to postmodern and poststructural thoughts and who inspired us to incorporate that thinking into our own research and practice.
- All of the students who we have had the privilege to work with and pass on what we have learned about postmodern and poststructural approaches.
- Colleagues near and far with whom we have engaged in largely discussions over the years related to this area
- Our Institutions for supporting our research and teaching in ways that enabled us to bring our postmodern and poststructural work together into a book that we hope will be useful and inspiring to colleagues and students as they continue their academic and practice journeys.
- Those who contributed to the first book and for those who have joined the conversation along the way between the two books.

We admire and thank those people whom we are close to and who have lived the ups and downs with us through the development of this book.

Preface

Nearly three decades ago Julianne wrote a book called *Postmodern and Poststructural approaches to Nursing research*. In the preface to that book she wrote:

> We hear and read the terms postmodern and poststructural with increasing frequency in the nursing and health care arenas. Yet their increased usage has not necessarily meant that these terms are well understood, nor that their potential for contributing to research endeavors in the health field is recognized. This was one of the reasons I decided to embark on writing a book such as this. I wanted to show that postmodern and poststructural approaches to research do not have to be obscure, ambiguous or poorly understood. Rather, I believe that some of the criticism of these approaches in terms of their inaccessibility and unintelligibility stems from a lack of definition or even at times the poor scholarship with which they are used and discussed.
>
> Secondly, I had experienced both rejection of papers I had prepared for journals, and difficulty in attracting funding for research using these approaches. This led me to question why this was so. What assumptions were being made about research and scholarship that seemed to preclude writing and researching within these frameworks? Was the rejection of a paper, for example, based on what it said, how it said it, what it didn't say and/or assumptions about what it should say? Much of the discussion to follow in this book shares with readers my exploration of such questions and where it has led me.
>
> The book makes no claim to providing definitive answers about either these issues or postmodern and poststructural approaches. Rather, it provides a beginning point from which readers can embark on their own journey of exploration of the potential afforded nursing and health care by these approaches.
>
> I have attempted to write the book in an accessible and user-friendly way. I hope that you, the reader, are able to engage with the material and that the further readings suggested assist you to develop your

understandings of postmodern and poststructural research approaches. Use the book as a "critical friend" in your journey of exploration.

(Cheek, 2000)

A lot has happened in the 20 years since Julianne wrote that preface and that book. Debates in practice areas continue about the meaning, purpose, and application of postmodern and poststructural thought. The ways in which this thought is discussed and debated remain intriguing and important to understand. Broader audiences have access to many of these ideas through podcasts, videos, and online articles and books. Online platforms and social media have added to the way we share information from both academic and political perspectives. There is a new generation of students, academics, and practitioners who are taking the ideas of postmodern and poststructural approaches and applying them to their own learning, research, and everyday practices.

Authors who have written recent articles and books about postmodern and poststructural approaches continue to discuss the importance of including a clear understanding of the foundational concepts (Aston, 2016; Fox, 2016; Mease, 2017; Ruitenberg, 2018). Further there have been many excellent books published on the philosophy and theory of postmodern and poststructural approaches.

However, there have been few books emerge that provide an accessible introduction to ways in which researchers in applied areas (for example health, education, and other professional practice areas) might draw on these perspectives to develop research able to make a difference to the practice worlds that they inhabit. Students and colleagues continue to remind us that, especially initially, the core concepts of postmodern and poststructural approaches can be challenging to understand.

Many of them have also told us that after reading Julianne's book they were better able to grasp and therefore navigate this complicated area when attempting to apply these theoretical ideas to research and practice. Given this, a few years ago, when talking with Julianne, Megan couldn't help but ask, when would she write the next updated version of her book as Megan and her students were looking for something more 'recent.' Over the course of a few meetings through Skype followed by a sabbatical visit to Norway, Julianne and Megan agreed that it would be a good idea to not just update Julianne's book but extend it beyond just a focus on nursing. This book is the result.

The updated book is designed for use by all students and researchers working in practice areas and who want to use postmodern and poststructural inspired approaches to make a difference to that practice. The purpose of the book is to provide readers new to these theoretical perspectives with an introduction to, and guide for navigating, this complex theoretical area. To do so we ground our discussion in examples of how postmodern and poststructural approaches have been, and can be, used in practice and applied areas. When doing so we provide tips about how to apply postmodern and poststructural thinking to practice in ways that can change that practice, by enabling us to understand how things have come to be the way they are and what sustains that status quo.

The updated book reflects what we have learned along the way on our journeys as educators, supervisors, and mentors for students and colleagues using these perspectives. We have supervised, worked with, and have seen Masters and PhD students use postmodern and poststructural approaches in ways that have shifted their thinking and the thinking of others to a place that provides critical social critiques that have made a difference in practice. The results have been exciting. Drawing on these experiences, we share interviews with colleagues and writings from students. We also provide activities for readers to undertake to help them further understand and implement these approaches in their own research endeavours.

In the 3 decades since the publication of the first book, we have seen changes in the use of, and reaction to, postmodern and poststructural approaches – both positive and negative. We intend for this book to be part of an ongoing dialogue, linking the past and present in a way that can pave the way for future students, academics, researchers, and practitioners to be comfortable and excited to use postmodern and poststructural concepts in their everyday practices.

What's in a Name?

We have chosen to use the phrases postmodern approach/thought and poststructural approach/thought rather than postmodernism and poststructuralism. We have purposefully decided not to use the words postmodernism and poststructuralism as this implies there might be one way to understand these approaches. In fact, different scholars have refused to be labelled postmodernists or poststructuralists, as this might limit the way people discuss postmodern and poststructural approaches. However, some of the references we refer to have used the term 'poststructuralism' and so we retain that term in those instances.

References

Aston, M. (2016). Teaching feminist poststructuralism. Founding scholars are still relevant today. *Creative Education*, *7*(15), 2251–2267. https://doi.org/10.4236/ce.2016.715220

Cheek, J. (2000). *Postmodern and Poststructural Approaches to Nursing Research*. Thousand Oaks, Sage.

Fox, N.J. (2016). Health sociology from post-structuralism to the new materialisms. *Health*, *20*(1), 62–74. https://doi.org/10.1177/1363459315615393

Mease, J.J. (2017). Postmodern/poststructural approaches. In Craig R. Scott & Laurie Lewis (Editors-in-Chief), James R. Barker, Joann Keyton, Timothy Kuhn, & Paaige K. Turner (Associate Editors), *The International Encyclopedia of Organizational Communication*. John Wiley & Sons. https://doi.org/10.1002/9781118955567.wbieoc167

Ruitenberg, C.W. (2018). Postmodernism and Poststructuralism. In P. Smeyers (Eds.), *International Handbook of Philosophy of Education*. Springer International Handbooks of Education. Cham, Springer. https://doi.org/10.1007/978-3-319-72761-5_51

1 Let the Journey Begin

Towards an Understanding of Postmodern and Poststructural Approaches

---

In this chapter, we will

- Introduce the idea of postmodern and poststructural approaches
- Highlight the importance of clearly articulating the understanding of postmodern and poststructural approaches being used in a research study
- Introduce how postmodern approaches question assumptions within modern thought
- Establish that poststructural approaches explore and analyse textual representations of reality
- Address questions about the legitimacy of postmodern and poststructural research approaches
- Consider how postmodern and poststructural research studies can focus on and offer unique research insights into practice

---

Postmodern and Poststructural Ways of Thinking

It is paradoxical that as the use of the terms *postmodern* and *poststructural* has increased across many areas such as health and other applied practice areas, so has the ambiguity and lack of clarity with respect to both these approaches. In some ways not much has changed since Watson (1995) declared nearly 3 decades ago: "just exactly what is postmodernism is unknown and ambiguous at best" (p. 60); supporting Bauman's earlier observation (1992) that postmodern thought was hard to define and did not represent a unified position.

Postmodern and poststructural approaches are more of "a set of intellectual propositions" (Bertens, 1995, p. 9) or philosophical positions that privilege "no single authority, method or paradigm" (Denzin & Lincoln,

DOI: 10.4324/9780429053764-1

1994, p. 15) that can be clearly delineated. Hence, postmodern and post-structural approaches are not research methods in themselves: rather they are ways of thinking about the world that shape the type of research that is done and the types of analyses that are utilized in that research.

Where to from Here?

Where, then, does this leave us? Is it the case that these terms are necessarily ambiguous and vague, unable to be defined? Or is it possible to attempt some clarity with respect to the debate about what these terms mean?

Perhaps a good place to begin is to ask why there continues to be such a lack of clarity. As we shall see throughout this book, much of the ambiguity arises from the different ways in which these terms are used by different scholars. What Daniel (1995) noted decades ago remains the case today, "there are so many different senses of postmodern (postmodernism or postmodernity) floating about today that no one description could respect all of the ways the term has been appropriated by theorists and commentators" (p. 256).

Part of the reason for this is that in many articles and books, the understanding of these approaches in use is not clearly articulated. In criticizing the ambiguity associated with the terms *postmodern* and *post-structural*, Bloland (1995) noted that it is often assumed that "those who use the words also know the theory" (p. 522). Therefore, too often, writers assume shared understandings of, rather than overtly stating, what they are postulating with respect to postmodern or poststructural approaches.

Of course, this criticism does not only apply to writers using postmodern or poststructural approaches in their work. How often have you read research reports where the approach used, such as "grounded theory," "phenomenology," "thematic analysis," or even "qualitative design," is stated as a given and no further explanation is deemed necessary? Yet within grounded theory, phenomenology, thematic analysis, and qualitative design there are a number of different perspectives each with its own emphasis and understandings. This is also the case with respect to postmodern and poststructural approaches, as these terms continue to get "stretched in all directions across different debates, different disciplinary and discursive boundaries" (Hebdige, 1988, p. 181).

Given this it is not possible to arrive at "*the*" or "*a*" definition for either one of the approaches. Indeed, the very nature of the approaches themselves militates against this, as we shall see in the discussion throughout the book. Therefore, if you were hoping that we could provide you with a neat one- to two-line definition for each term here in the opening sections of this book, you will be disappointed and we make no apology for that!

However, *contestability of meaning* is not necessarily synonymous with *lack of clarity*. It is possible to explore the reasons for such contestability and then to arrive at an informed and stated position for any piece of work.

Any research or discussion which purports to draw on postmodern or poststructural approaches must clearly articulate the understanding of postmodern or poststructural approaches that underpins that study or that research. While not all may agree with such a stated position, at least there will be a position and it will be possible to determine what the position is. It is in this spirit that this introductory chapter, and in fact the entire book, is written.

Key point

Simply stating terms such as "postmodern," "phenomenology," "grounded theory," and so on is not enough. The reason for choosing a particular approach to frame a particular piece of research and the analysis arising from it must be stated, as must the particular understanding of the approach being employed.

Introducing Postmodern Approaches

All approaches and propositions considered as postmodern question the assumptions embedded within modernist thought. Indeed, postmodern thought has been described as a "crisis of confidence in the narratives of truth, science and progress that epitomized modernity" (Bauman, 1992, p. 98). Lyotard referred to this as "a postmodern condition" (1984) which was "marked by the gradual erosion of the certainties of modernity founded in science, rationalism and humanism by a perfect storm of the information revolution, post-industrialisation, pluralism and academic relativism" (Fox, 2016, p. 65).

The emphasis in modern theoretical analysis is on the big picture; that is, grand theories of social structure and action. In contrast postmodern thought disavows the idea that human experience can be reduced to and captured by grand or totalizing theories. Rather, postmodern thought emphasizes the plural nature of reality, the multiple positions from which it is possible to view any aspect of reality including health care, and the partial nature of any representation of reality that arises from any form of writing/speaking that attempts to explore, describe, or explain that reality. This is why in later work Bauman (2012) proposed the idea of 'liquid modernity' in which 'uncertainty was the only certainty' (explored further in Chapter 2). Such liquidity "parallels the global tendency to move away from univocal, modernist perspectives to the more complex, and ambiguous postmodern ones" (Chastagner, 2022, p. 90).

There are numerous examples that can be found today of scholars in many different fields who critically apply such a postmodern approach. For example, Hasanova (2023), when commenting on the work of author David Mitchell and the importance of using postmodern techniques in books of fiction states "Mitchell often used multiple voices and perspectives to tell his stories...In general, it can be said that David Mitchell is a prime example of a postmodern writer..." (p. 83).

Postmodern approaches can thus be described, at least in part, as a response to what has come to be viewed as a crisis in representation – a challenge to the view that it is possible to represent reality, speak for others, make truth claims, and attain universal essential understandings. As Koro-Ljungberg, Douglas, Carlson, and Therriault (2015), drawing on Frank (2013) point out,

> when we limit who we see as a competent knower we bound our ability to gain a full understanding of our complex world and experiences. In addition, by narrowly defining what counts as "truth" or by using the "right method" as the fixed standard, scholarly communities might severely bound their ability to see diverse perspectives and support diversified and less familiar practices.
>
> (p. 48)

Postmodern approaches recognize the presence of multiple voices, multiple views and multiple methods when analysing any aspect of reality. Who and what is absent from representations of that reality is thus of as much interest as who or what is *present*. All of this challenges the notion of a rational and unified subject that is so central to modernist thought.

What is taken as being natural or normal, that is "a given" or "truth," is open to question and challenge in postmodern approaches. This includes the everyday practice settings of practitioners in areas such as health, education, and business. For example, instead of seeking to describe and understand how a particular practice setting functions (e.g. the classroom or the clinic), postmodern approaches allow for the possibility of exploring how the practice setting *itself* came to be constructed in the way that it is, what enables it to stay that way, and how it might be different. What are the assumptions and understandings of practice that are taken for granted and which have shaped the way practice settings operate? Whose assumptions and understandings are they, and why are other views excluded or marginalized?

Rounding Off

Based on the preceding introductory discussion a working definition of postmodern approaches is that postmodern approaches reject "modern assumptions of social coherence and notions of causality in favour of multiplicity, plurality, fragmentation, and indeterminacy" (Best & Kellner, 1991, p. 4).

This is where we will leave our introduction to what postmodern approaches are about. This introductory discussion is picked up on and developed in Chapter 2 which explores postmodern thought, what it is, and how it can be applied to applied practice areas, in more detail.

Introducing Poststructural Approaches

Poststructural approaches have much in common with postmodern approaches: in fact some writers have used the terms interchangeably. How aspects of reality such as practice settings (e.g. the health clinic or the school classroom) and the practices in them, are constructed and maintained is central to both postmodern and poststructural analyses. However, the analyses differ in terms of their focus and emphasis.

Poststructural approaches tend to focus on the exploration and analysis of *texts* where texts refer to representations of reality.

> With their focus upon texts and textuality, post-structuralists rejected sociological conceptions of an over-arching social structure, and sought instead to explore how systems of thought or 'discourses' shaped social action in ways that were historically and culturally contingent, and thereby only ever partial versions of a phenomenon."
>
> (Fox, 2016, pp. 63–64)

For example, what are the assumptions being made in texts about how things should/must be in a particular context or setting; whose assumptions are they; who and what is excluded or relegated to the margins by this view of reality and normality? Texts such as patient case notes and the form they take, classroom procedure manuals, observations, visual records (e.g. photographs) of what happens in a practice setting, or the transcript of an interview between parents and teachers about a 'problem' child.

Squier (1993) neatly encapsulates what poststructural inspired research is about and when doing so provides a working understanding of poststructural approaches. She writes that the task of poststructural research approaches is to "investigate the meaning of particular representations: to understand how they came to be as they are, and what they communicate about their specific cultural and historical contexts" (p. 30). Harcourt (2007) agrees and writes "...structures of meanings are not universal, and

do not reflect ontological truths about humans or society. Poststructuralists focus on those gaps and ambiguities in the system of meaning and find
meaning there" (p. 17).

The following are examples of some studies that use poststructural
thought to "investigate the meaning of particular representations:
to understand how they came to be as they are, and what they communicate about their specific cultural and historical contexts"
(Squier 1993, p. 30).

1 Cheek's (1997) analysis of the way that Toxic Shock Syndrome (TSS) was talked and written about in the media and
how this promoted certain understandings of TSS while at
the same time limiting the possibility of alternative views of
TSS being presented. An extended discussion of this study
can be found in Chapter 6.
2 Gee and Skovdal's (2018) study of the stigmatization of international health workers after they returned to their home
countries after working in West Africa during the Ebola outbreak demonstrates how media reports promoted and constructed certain understandings of both risk and the risk these
returning workers posed to their communities. At the same
time other understandings (such as the actual infectious state
of these workers – which was minimal or most often zero) were
relegated to the margins. Such media derived texts resulted in
"fearful social representations of the disease and public discourses of contagion and blame led to the returning health
workers to become tangible embodiments of the disease in
their home countries, regardless of actual infectious status."

(Gee & Skovdal, 2018, p. 1505)

We will leave our introduction to the idea of what poststructural
approaches are here. This introductory discussion is picked up on, and
developed, in Chapter 4 which explores poststructural thought, and how
it can be applied to practice areas in more detail.

A Word of Caution

A word of caution is necessary before we leave this introduction to postmodern and poststructural approaches. Although for the purposes of
convenience we have divided the structure of the book into individual

chapters about postmodern approaches and poststructural approaches, this is not to imply that such a neat division is without problems. We have already stated that these approaches have much ground in common: with both valuing plurality of thought and perspective and challenging taken for granted aspects of reality, including practice areas.

It is important to keep in mind what Agger (1992) declares about these approaches, "at some level, they are so inextricably linked as to make simplistic differentiations impossible or undesirable" (p. 109). Indeed, Daniel (1995) argues that poststructuralism, like critical theory and deconstruction, needs to be "situated within a broadly postmodern context" (p. 257). Likewise, Best and Kellner (1991) argue that a poststructural approach forms part of "the matrix of postmodern theory" and thus a subset of a broader approach (p. 25).

Consequently, it is not possible to clearly separate postmodern thought from poststructural thought. This is evident from the way in which various theorists are described as 'poststructural' by some scholars and 'postmodern' by others. For example, while Derrida's work is usually classified as poststructural (although he does not identify himself as such), Foucault's work has been described as both poststructural and postmodern. Like the work of these prominent theorists, postmodern and poststructural approaches resist being placed into neat, clearly delineated categories. However, while there are many similarities between these two approaches, as we have seen, each does have its own specific emphases (Daniel, 1995) and levels of focus. We will develop and explore this point further in Chapters 2 to 5.

Can Postmodern and Poststructuralist Approaches Influence Practice or Are They Merely Esoteric?

A key thread running throughout the discussion in the book is the question of whether or not postmodern and poststructural approaches are able to contribute to understandings of practical aspects of health, education, or any other type of practice or applied setting.

Some critics claim that these approaches are merely esoteric and inward looking, unable to make any contribution to practice areas at all. Indeed, they go on to suggest that because "postmodernist constructions of knowledge are necessarily local, contextual and readily contestable, their social significance is minuscule" (Kermode & Brown, 1996, p. 383). Or put another way, drawing on Popper's (1976) examination of critical theory, such critics ask whether these approaches are anything more than trivialities in high-sounding language. If this is the case, then we should abandon our discussion at this point, for what is the value of such an approach to research in practice areas, if the product of that research cannot contribute to the improvement and development of practices in those areas? The answer, obviously, is very little value, if any.

However, what these types of critics *don't* say is just as important as what they *do*. Such critics approach their own research *and* base their critique and judgments about the value (or lack of it) of certain types of research from a particular world view. As a result of these (often undeclared) background assumptions (Gouldner, 1971) about what is and is not research, not all research approaches are afforded equal credibility and standing. Often legitimacy of a piece of research is conferred by certain groups who have a certain view about what constitutes "valid" research and "valid" research methods. This highlights Chia's (1996) assertion "that the researcher/theorist [critic] plays an active role in constructing the very reality he/she [sic] is attempting to investigate (or in this case critique)" (p. 42).

The position from which commentators take up their critique of postmodern and poststructural approaches is an important issue, but it is often ignored especially with respect to discussions about the value of these approaches in terms of their ability to influence or change practice. Mullhall (1995) neatly encapsulates the central point we are making when she states that "the attachment to traditional and already accepted concepts and paradigms may also overwhelm new evidence presenting a different view... certain people define phenomena in certain ways, and once defined there is considerable resistance to change" (p. 580).

The Effect of Funding on How Legitimacy and Utility of Research Is Viewed

In the climate of economic rationalism and the marketization of higher education institutions cost benefits have become a prime focus for evaluating the value of a certain health care procedure or teaching method or other type of practice (Cheek, 2018, 2024). The same is true for research. This is particularly so as universities are increasingly urged to "pay their way," so that funding for research rather than the quality of the research itself often becomes the hallmark of success. (Cheek, 2022, 2024).

Funding bodies have specific understandings, often not overtly declared of what research is and what "good" research is about. "This is in relation to both the substantive focus of the proposed research – the problems, or areas of a problem, that will be prioritized – **and the *type* of knowledge about these problems/areas that is being sought**" (bold is our addition) (Cheek, 2018, p. 339).

One key consideration for many funding bodies is the immediacy of the effects of the recommendations of the research undertaken. In such a view, research must be able to immediately and directly affect practice or be applied by enabling the swift implementation of recommendations. Funding bodies may even commission research the aim of which is to find solutions to the problems that the funder has identified as problems (Cheek, 2022).

What we are really talking about here is the nature of the research product or, to use a term that is achieving much prominence in contemporary research rhetoric, the 'deliverable(s).' Research deliverables are increasingly being defined as recommendations, strategies, and models for best practice and/or cost efficiency in health service delivery. There is no doubt that such deliverables can have a positive impact on practices and the content and delivery of services and programmes in areas such as health care and education, and thus should be a focus of applied and practice-based research. However, they should not be the *only* focus. There is room for other research approaches and outcomes that can inform understandings of practice itself. If we are only interested in improving what is, it may well be that we will never explore what might be and actually be a better option.

Indeed, even strident critics of postmodern and poststructural approaches such as Kermode and Brown (1996) are grudgingly forced to concede that "as methods and procedures for gathering and interpreting data, postmodernist approaches may produce a useful starting point in some areas" (p. 383). They are a way of combatting a "stable conservatism" (Foster et al., 2015, p. 900) reflected in the types of areas and issues being funded (Cheek, 2018).

Therefore, while postmodern and poststructural research approaches may not be able to produce immediate cost-benefit analysis on their own, they may be able to provide the basis for conceptualizing practice in new or different ways which can then produce better educational or health practices. Thus, although research from these approaches may be in this sense "once removed" in terms of how immediate the applicability of the findings are, this type of research can influence, inform and improve basic understandings of what the problem actually is (Bacchi, 2012).

To illustrate this point, consider the outcomes of the study Julianne did called "Constructing Toxic Shock Syndrome: Selected Australian Print-Based Media Representations of Toxic Shock Syndrome from 1979–1995" (see Chapter 6). This research explored the discursive construction of a relatively new health phenomenon, Toxic Shock Syndrome (TSS) by Australian print-based media. In so doing, it aimed to provide insights into ways print-based texts represented health issues, thereby influencing attitudes towards health, towards those with certain illnesses and towards health care practices. Thus, in the section of the proposal asking for direct applications of research results, it was possible to make the following points. The study would:

- provide better understanding of the way in which "popular" magazines (a largely neglected research focus) and newspapers frame and represent health issues and risk with respect to TSS, and
- explore ways in which these print-based media might better communicate understandings of health and health risks.

It was then possible to point out that once such representation was analysed, and its effects better understood, it would be possible to design relevant health promotion strategies and information packages pertaining to
TSS, the use of tampons, and other related health issues for both consumers and health professionals. The understandings provided by this study
had the potential to inform and influence the delivery of health care and
attendant health care practices.

All of this is to recognize that the criticism that is often levelled at
postmodern, poststructural, and other research approaches about their
apparent inability to contribute to practice may well be more a political
criticism than a methodological one. It is a criticism premised on a certain
understanding of what a research deliverable or product is, or should be,
particularly in terms of its perceived immediate instrumental applicability
(Cheek, 2022, 2024).

The rest of this book will demonstrate that to ignore such approaches
is to forfeit the potential for practice-oriented research to evolve and
develop multiple research approaches from which to explore the reality of
that practice, and thereby open up possibilities for real change in those
applied settings and areas. Practical, specific, and concrete research outcomes are needed in practice-based areas and disciplines such as health
and education, but so are thoughtful practitioners who can influence and
change practice. The two need not be mutually exclusive.

What Is to Follow?

This chapter has set the parameters for the way in which this book is organized, what it aims to achieve, and the major ideas that form the focus of
the discussion to follow. These points are amplified in Chapters 2 to 5.

Chapter 2 is specifically concerned with *postmodern* thought. It provides an overview of this type of thought and explores why it is often
viewed as ambiguous and hard to define. Rather than providing "the"
definition of a postmodern approach, the chapter situates postmodern
thought theoretically, and thereby adds depth to a working definition of a
postmodern approach.

Chapter 3 consolidates and builds on the discussion in Chapter 2 and
considers the work of one of the most influential theorists whose work
has been associated with postmodern/poststructural thought and perspectives—Michel Foucault. Examples of how researchers have used Foucauldian analysis to inform our understandings of aspects of practice are
provided and discussed. In particular, Foucault's notion of *discourse* is
explored.

Chapter 4 turns the reader's attention to *poststructural* thought. A key
point is made at the outset (which we have already touched briefly on in
this chapter), namely that *poststructural* and *postmodern* are in many ways

closely related terms and are often used synonymously. However, our exploration develops the point made earlier in this chapter that poststructural analyses are often concerned with an analysis of texts that represent aspects of reality. Deconstruction as an approach is discussed.

Chapter 5 explores discourse analysis as a research approach and how it draws upon poststructural approaches. We provide examples of how to put discourse analysis into practice.

The remaining chapters of the book focus more on "how to" aspects of employing postmodern and poststructural approaches when thinking about and researching applied practice areas.

Chapter 6 explores a research proposal written by Julianne guided by postmodern and poststructural approaches. Areas explored include, writing for a particular audience, defining the issue/question of what constitutes a topic that can be researched, using the literature to help define the research issue/question, using the theoretical perspective drawn upon to inform the method, and specifying practical outcomes and benefits of the research in terms of health care practice.

Chapter 7 continues the discussion regarding how to write a proposal using these approaches. A research proposal written by Megan and a colleague is used as the vehicle for the discussion as are reviewers' comments on the proposal and the way that these reviewer comments were addressed. The chapter includes tips to help readers develop their own research proposal.

Chapter 8 draws together the threads of our book length discussion of how we might use postmodern and poststructural approaches to practice. It includes reflections on the use of these approaches on practices including some by students as part of their graduate studies. The chapter ends by encouraging readers to be self-reflexive when using these approaches in their own research.

Coda: In this section we provide examples of researchers' personal experiences of, and perceptions related to, using postmodern and poststructural perspectives in their research. These are insightful reflections on different aspects of the challenges, but also the rewards of engaging with such perspectives. We call this section a Coda as a Coda in music is an ending that concludes and rounds out in a different way or form what has gone before it.

References

Agger, B. (1992). *Cultural Studies as Critical Theory*. London, Falmer Press.

Bacchi, C. (2012). Why study problematizations? Making politics visible. *Open Journal of Political Science, 2*(1), 1–8.

Bauman, Z. (1992). *Intimations of Postmodernity*. London, Routledge.

Bauman, Z. (2012). *Liquid Modernity*. Polity Press.

Bertens, H. (1995). *The Idea of the Postmodern: A History*. London, Routledge.

Best, S., & Kellner, D. (1991) *Postmodern Theory: Critical Interrogations*, New York, Guilford Press.

Bloland, H. (1995). Postmodernism and higher education. *Journal of Higher Education*, *66*(5), 521–559.

Chastagner, C. (2022). Postmodern intercultural communication: Beyond national and ethnic identities. *Culture Studies*, *18*(1)(25), 90–100. https://doi.org/10.46991/AFA/2022.18.1.090

Cheek, J. (1997). (Con)textualizing toxic shock syndrome: Selected media representations of an emergent health phenomenon 1979–1995. *Health*, *1*(2), 183–203.

Cheek, J. (2018). The marketisation of Research: Implications for qualitative inquiry. In N.K. Denzin, & Y.S. Lincoln (Eds.), *The Sage Handbook of Qualitative Research* 5th edition. Sage Publications, pp. 322–340.

Cheek, J. (2022). The impact of funding on ways qualitative research is thought about and designed. In U. Flick (Ed.), *The SAGE Handbook of Qualitative Research Design* 2 Volume, pp. 636–651.

Cheek, J. (2024). Academic Survival: Qualitative Researchers in the Neo-liberal Academy. In N.K. Denzin, Y.S. Lincoln, M.D. Giardina & G.S. Cannella (Eds.), *The SAGE Handbook of Qualitative Research* 6th edition. pp. 599–615.

Chia, R. (1996). The problem of reflexivity in organizational research: Towards a postmodern science of organization. *Organization*, *3*(1), 31–59.

Daniel, S. (1995). Postmodernity, poststructuralism, and the historiography of modem philosophy. *International Philosophical Quarterly, XXXV, 3*(139), 255–267.

Denzin, N., & Lincoln, Y. (1994). Entering the field of qualitative research. In N. Denzin & Y. Lincoln (Eds.), *Handbook of Qualitative Research*, California, Sage, pp. 1–18.

Foster, J.G., Rzhetsky, A., & Evans, J.A. (2015). Tradition and innovation in scientists' research strategies. *American Sociological Review*, *80*(5), 875–908.

Fox, N.J. (2016). Health sociology from post-structuralism to the new materialisms. *Health*, *20*(1), 62–74. https://doi.org/10.1177/1363459315615393

Frank, J. (2013). Mitigating against epistemic injustice in educational research. *Educational Researcher*, *42*(7), 363–370.

Gee, S., & Skovdal, M. (2018). Public discourses of Ebola contagion and courtesy stigma: The real risk to international health care workers returning home from the West Africa Ebola outbreak? *Qualitative Health Research*, *28*(9), 1499–1508.

Gouldner, A. (1971). *The Coming Crisis of Western Sociology*. London, Heinemann.

Harcourt, B.E. (2007). An answer to the question: 'What is poststructuralism?' University of Chicago, Public Law Working Paper, (156).

Hasanova, R.S. (2023). The manifestation of postmodern traditions I the work of David Mitchell. *International Journal of Literature and Languages*, *3*(4), 82–87. https://doi.org/10.37547/ijll/Volume03Issue04-15

Hebdige, D. (1988). *Hiding in the Light: On Images and Things*. London, Comedia.

Kermode, S., & Brown, C. (1996). The postmodern hoax and its effects on nursing. *International Journal of Nursing Studies*, *33*(4), 375–384.

Koro-Ljungberg, M., Douglas, E.P., Carlson, D., & Therriault, D.J. (2015). An unfinished dialogue about problematising knowledge production in the peer review process. In N.K. Denzin & M.D. Giardina (Eds.), *Qualitative Inquiry and the Politics of Research*, Left Coast Press, pp. 27–50.

Lyotard, J. (1984). *The Postmodern Condition: A Report on Knowledge*. Minneapolis, University of Minnesota Press.

Mullhall, A. (1995). Nursing research: What difference does it make? *Journal of Advanced Nursing, 21*(3), 576–583.

Popper, K. (1976). Reason or revolution. In R. Adorno, R. Dahrendorf, J. Habermas, H. Pilot, & K. Popper (Eds.), *The Positivist Dispute in German Sociology*, London, Heinemann, pp. 288–300.

Squier, S. (1993). Representing the reproductive body. *Meridian, 12*(1), 29–45.

Watson, J. (1995). Postmodernism and knowledge development in nursing. *Nursing Science Quarterly, 8*(2), 60–64.

2 Postmodern Thought and Its Possibility to Influence Practice

---

In this chapter we will

- Establish both original and present-day understandings of post-modern thought
- Examine various struggles and critiques faced by researchers and academics when applying this thought to practice areas.
- Show how postmodern thought has responded to, and developed in light of such critiques
- Offer a beginning working understanding of postmodern thought
- Demonstrate how postmodern thought can inform practice
- Analyse challenges to postmodern thought today
- Explore what research using a postmodern approach might actually look at, what it might involve, and provide examples of such research

---

Postmodern Approaches: A Way of Thinking

Postmodern thought, and debates associated with that thought, have gained increasing import and interest in many disciplinary fields since World War II. The development of such thought has been strongly influenced by French theorists such as Foucault, Lyotard, and Baudrillard to name a few. Although, as stated in Chapter 1, many of these theorists rejected the placement of their work into any particular category—including that of 'postmodernism.' Understandings of postmodern approaches today continue to support many of the original concepts proposed by these theorists over 40 years ago.

Initially postmodern thought influenced predominantly the fields of art and architecture and then spread to philosophy and literary studies in the 1950s through to the 1990s. Since then, postmodern thought has continued to impact, and be used, in an increasing number of fields of study

DOI: 10.4324/9780429053764-2

including practice areas such as health care and education. Seen as a movement as well as a philosophy or theory, postmodern approaches have developed into "a broad movement encompassing a range of fields and disciplines beyond linguistics including philosophy, politics, architecture, art, literature and the social sciences" (Baxter, 2016, p. 35).

Best and Kellner in their book "Postmodern Theory: Critical Interrogations" (1991) provide a succinct yet comprehensive overview of what they term the archaeology of the postmodern. If you want to read more about the influence of particular theorists and schools of thought on the early development of understandings of postmodern approaches this is an excellent place to start.

Postmodern Approaches – An Unstable and Contested Concept

Postmodern thought represents an unstable concept (Bauman, 1992) that is hard to define in that it does not represent a unified position or a coherent school of thought. For example, there is diversity of thought as to what the prefix post refers to in the term postmodern. Some writers view the prefix as a "periodizing concept whose function is to correlate the emergence of new features in culture" (Sarup, 1989, p. 131) where post refers to the next step in the historical development of ideas and cultural features in contemporary Western society. In this sense postmodern is post in that it follows and succeeds the modern era.

Others view the prefix post as signifying "an active rupture (coupure) with what preceded it" (Best & Kellner, 1991, p. 29) – namely modernity and modern thought. While others such as Holtzhausen (2013) view postmodern thought as an outflow of modernity.

> I wish to argue that postmodernity in many instances is an outflow of modernity, rather than a rupture with modernity; that pitting the modern against the postmodern is a form of intellectual blackmail that forces one to choose between the two.
>
> (p. 3)

However, while there is such variation related to the term "post" this is not to suggest that there is nothing in common among these positions and approaches termed postmodern. All postmodern approaches emerge from a critique of the assumptions that underpin modernist thought. Indeed Bauman (1992) declared that it is the dismantling of the artifice of modernity, and all that that artifice is premised on, that is the focus of postmodern thought. Similarly, Bertens (2003) when writing about the different thinking about, and approaches to, postmodern thought noted that "In their own way, they all seek to transcend what they see as the self-imposed limitations of modernism, which in its search for autonomy and purity or for timeless, representational, truth has subjected experience to unacceptable intellectualizations and reductions" (p. 5).

Thus, while there is variation in the way that "post" is thought of, generally most scholars would agree with Mease (2017) who writes

> The 'post' attached to structuralism and modernism signals both a response and a critique. 'Post' movements are not so much absolute rejections of, but extensions born out of the failing of structural and modern attempts to adequately describe the world.
>
> (pp. 1–2)

As a way of navigating such different thinking about the idea of post – Bauman (2012) challenges definitive interpretations of 'post' and has proposed the idea of 'liquid modernity.' He writes

> Being always, at any stage and at all times, 'post-something' is also an undetachable feature of modernity. As time flows on, 'modernity' changes its forms in the manner of the legendary Proteus... What was some time ago dubbed (erroneously) 'post-modernity,' and what I've chosen to call, more to the point, 'liquid modernity,' is the growing conviction that change is the only permanence, and uncertainty the only certainty.
>
> (Foreword p. 1)

With all of this in mind we are now (and only now) ready to develop a working understanding of a postmodern approach that will be used throughout the book. In the next section we work towards that.

Towards a Working Understanding of a Postmodern Approach

As we have seen, postmodern thought problematizes modernist understandings of history as having a definite direction which is ultimately progressive, the desire for universal and generalizable categories of explanation, a belief in reason as the basis for action and a belief that "the nation state could coordinate and advance such developments for the whole society and thereby constitute society itself" (Parton, 1994, p. 101).

Postmodern thought discards the organic notion of all parts of society working together in an orderly way, and in so doing rejects "modern assumptions of social coherence and notions of causality in favour of multiplicity, plurality, fragmentation, and indeterminacy" (Best & Kellner, 1991, p. 4).

Postmodern thought thus rejects grand theories which aim to offer totalizing descriptions and explanations of both history and social structures. Indeed, Lyotard (1984), a foundational French postmodern writer, in his seminal work 'The Postmodern Condition: A Report on Knowledge," encapsulates his understanding of postmodern thought in the following statement: "Simplifying to the extreme, I define postmodern as incredulity towards metanarratives" (p. xxiv). By metanarratives, Lyotard is referring to conceptual and theoretical schema that attempt to link and represent all aspects of reality coherently in a way that is supposedly consistent and true.

Agger (1992) captures this idea neatly when he states that a postmodern approach "is a theory of cultural, intellectual and societal discontinuity that rejects the linearism of Enlightenment notions of progress" (p. 93). Rather than seeking universal and essential truths, postmodern thought recognizes the existence of multiple perspectives, assuming instead plurality of understandings for any aspect of social reality. For example, postmodern thought challenges the very persuasive metaphor of the health care system as a system where all components and players work together for a common good. It also problematizes the notion of 'advances' and 'progress' in health care both in terms of what actually constitutes progress and advances, and concomitantly, whether developments and change in the health care system are necessarily progressive.

Mease (2017) supports many of the points made by these scholars from the 1990s when writing about postmodern approaches:

- Their "premises lead to the conclusion that social structures – and the truth claims embedded in them – are created through human interaction (including scientific inquiry) rather than discovered through scientific inquiry.
- This ontological shift has epistemological and methodological implications for postmodern approaches.
- If social structures are contingent upon human interaction, then claims about social structures must be grounded in and limited to a particular social and historical context, not universally applied across time and space.
- Thus, postmodern approaches that often emphasize ruptures, disjunctions, tensions, instabilities, and other inconsistencies demonstrate a methodological choice aimed at revealing the faulty, constructed, and precarious character of structures as creations.

- Because social structures are considered created rather than natural, postmodernism often demonstrates irreverence for commonly accepted norms or truth by playing with traditional rules or expectations."

(Para 10) (dot points added)

Note we have broken up Mease's original text using dot points to highlight each of the points Mease is making – all of which are interconnected and therefore no one dot point should be read in isolation. Mease's statement is a good example of a working understanding of postmodern thought and is the understanding that we will have in mind when we use the term throughout this book.

This understanding highlights that far from being nihilistic and only destructive, postmodern thinking is enabling and can offer constructive and useful analyses for practice. It is not simply an approach that allows "you to drive a coach and horses through everybody else's beliefs while not saddling you with the inconvenience of having to adopt any yourself" (Eagleton, 1983, p. 144) – a critique that four decades later still is often levelled at postmodern thought.

Not just nihilistic and only destructive

Murali (2019) discusses the usefulness of using a postmodern philosophical approach to help guide nurses in practice as they assist patients with end-of-life (EOL) decision-making. A postmodern perspective helps to understand the multiple and variable meanings of EOL decision-making. In the paper, Murali demonstrates how

> Postmodernism holds central the concept of subjectivity. It allows for clinicians and researchers to approach the phenomenon of EOL decision-making from multiple vantage points and accounts for all of the situational and contextual factors that may influence decisions.

(Para 4)

You can read this study in full:

Murali, K.P. (2019). End of life decision-making: Watson's theory of human caring. Nursing Science Quarterly. 33(1). https://doi.org/10.1177/0894318419881807

In the next section of the chapter we develop further the idea of post-modern thought as enabling as it encourages us to think about our socially constructed reality(ies) in reflexive and different ways.

Postmodern Thought as Enabling

Far from being destructive, postmodern thought enables a reflexivity that unmasks complex political/ideological assumptions often hidden in our practice and also the way we write about that practice. Defining reflexivity is not an easy task – like postmodern thought there "are numerous definitions and operationalizations of reflexivity" (Lumsden, 2019, p. 2). By reflexivity we mean "a type of folding or bending back (Finlay and Gough, 2003) on our own thinking to work out why we have come to think about something in the way that we do. What assumptions do we make when we think in this way? Based on what?" (Cheek and Øby, 2023 p. 18).

Such self-questioning encourages, and enables, us to think about the socially constructed reality(ies) which we are part of, and in part constructed by, in new and different ways. Truth claims are less easily validated now: "desire to speak 'for' others are suspect" (Richardson 1994, p. 523). This can be positive and constructive (Vattimo 2007) as challenging assumptions that have achieved truth status in, for example, practice areas, rather than just validating them opens up and presents multiple possibilities for change. Therefore, as much as postmodern thought is 'undetermined,' it is also 'undetermining' in that it serves to "weaken... the constraining impact of the past and effectively prevent... colonization of the future" (Bauman 1992, p. 190).

In a book edited by Scheurich (2014) *Research Method in the Post-modern*, authors write about the ethics and methods that researchers need to consider when conducting postmodern qualitative research including the importance of attending to self-reflexivity on personal location, language, meaning, and power.

Yet this is not to suggest that postmodern thought attacks all reason and discounts rationality in order to glorify irrationality (Daniel, 1995). Rather, it is to highlight the constructed nature of what have often become "truths" and taken-for-granted aspects of our reality. This includes practices and procedures in practice. It is important to recognize that:

> because postmodernity undermines the fascination of truth itself, it does not offer itself as the truth. Instead...it is an invitation to consider

how things would be different if we were to adopt such beliefs, an invitation to imagine what would happen if we were to think this or that way.

(Daniel, 1995, p. 266)

Aylesworth (2015) concurs with Daniel and writes,

That postmodernism is indefinable is a truism. However, it can be described as a set of critical, strategic and rhetorical practices employing concepts such as difference, repetition, the trace, the simulacrum, and hyperreality to destabilize other concepts such as presence, identity, historical progress, epistemic certainty, and the univocity of meaning.

(p. 1)

How We Can Use Postmodern Thought in Everyday Practices

The following example demonstrates how postmodern thought can provide a way to question and critique taken for granted everyday practices within the health care system. This then enables us to understand how different beliefs influence these practices and vice versa.

Julianne was interested in how health and health care were understood and enacted at any point in time. This interest was premised in her view of health care as

a space occupied, shaped and colonized by a variety of players at a variety of times...Players move in and out of this space (or are allowed in or moved out) all the time...This is "a contested and troubled space, one that is increasingly uncertain and ambiguous."

(Cheek, 2008, p. 974)

She wanted to explore how in this uncertain and ambiguous space healthism – "a particular way of viewing the health problem" (Crawford, 1980, p. 365) took on various guises and forms. When doing so Julianne declared that she "deliberately sought to destabilize aspects of the present that we find ourselves in, to challenge what otherwise might become the obvious, the ordinary, the normal and taken for granted, and the seemingly inevitable." (p. 981) Drawing on Foucault (1988), she "employed a spirit of critique" (p. 981) to highlight the importance of questioning: "What kinds of assumptions, what kind of familiar, unchallenged, unconsidered modes of thought might impact the way we think and speak about our research, and health care practice." (p. 981)

Using three examples of new forms of healthism "where boundaries of health and health care are being pushed, pulled and reshaped" (p. 975). Julianne explored how each of these examples was "underpinned by new understandings of old problems, such as how to avoid death, how to view and respond to risk, and how to remain in an ever - vigilant state – a new and transformed version of a "what if" approach to health, rather than a "what is" (Cheek, 2008, p. 974).

She explored how these new forms of healthism had refracted understandings of what being healthy meant by extending these understandings to "embracing a range of lifestyle choices and technologies... that once would have been considered at the periphery of health, if indeed part of it at all" (Cheek, 2008, p. 975). Consequently

> working out at a gym, or undergoing cosmetic surgery to feel good about ourselves, or taking drugs to increase sex drive...have become as much a part of healthcare practices as vaccinations, prescriptions for medicines, and taking our temperature. In such a climate, hitherto traditional boundaries such as where and how health is enacted and what healthcare actually is, dissolve and undergo constant change.
>
> (p. 975)

She was interested in "what this might mean for individuals receiving the services and those providing and/or researching those services" (p. 974).

She concluded that contemporary forms of healthism may promote a new conservatism where "health takes on new and different forms of discipline" (p. 980) – including increasing surveillance, both by the self and others.

> It is no longer enough to be without, or actively working to prevent, physical disease to be considered healthy. Health has become the new fountain of youth...a new version of the eternal quest for immortality, and a new form of a badge of honor by which we can claim to be responsible and worthy...in many Western contemporary societies health approaches sacred status: Healthism is to the fore.
>
> (p. 974)

You can read the full article at Cheek, J (2008) Healthism: A New Conservatism? Qualitative Health Research Vol 18 no 7 974–982.

Challenges to Postmodern Approaches Today

If postmodern thought is "an invitation to consider how things could/ might be different ... an invitation to imagine what would happen if we were to think this or that way" (Daniel, 1995, p. 266) it is not hard to see why some scholars might experience discomfort and misunderstandings regarding postmodern thought.

For example, there are recent video examples posted on YouTube where scholars attempt to discredit and de-legitimate postmodern approaches. However, when watching the videos it is often evident that they themselves do not understand postmodern ideas such as truth, multiple positions, or deconstruction.

For example, misconceptions presented in the following video clip include statements such as 'postmodernism does not believe in truth' https://www.youtube.com/watch?v=ynMS0g6zu5E&t=20s (Niles, 2010). As we have discussed in earlier parts of the chapter this is incorrect. Rather postmodern approaches claim there is no *one* truth. Rather there are multiple ways of understanding the world.

Other critics misrepresent this idea of no *one* truth to claim and instead there are multiple ways of understanding the world. For example, some claim that there is a "postmodern assault on science" and that "If all truths are equal, who cares what science has to say" (Kuntz, 2012, p. 885). That of course is *not* what postmodern thinking claims. It *does* however, challenge notions of only one way to do research and or understand practice. It also exposes vested interests in maintaining that particular view of research and/or practice. Susen (2015) believes there is a political and social fear from those who position themselves in a different research paradigm. While, there should be scholarly debates about different research paradigms; one would expect that debates are grounded in evidence and respectful dialogue.

Critics of postmodern thought also continue to question the relevance of its use in academia and practice, assuming it was only relevant when scholars used it to challenge political and social issues during a time that was labelled 'modernist.' However, there are many examples of how researchers, scholars and practitioners use postmodern approaches today. For example, Bertens (2012) comments on the relevance of using a postmodern approach. "We have moved on, but for many of the most interesting writers of our age the insights of postmodernism and the technical innovations of postmodern writing have lost nothing of their interest and relevance" (p. 315). Throughout the book we provide numerous examples of recent postmodern and poststructural research studies which have lost nothing of their interest and relevance.

Thus, although there are critics who continue to challenge postmodern thought to be amoral, disruptive, confusing, and even redundant, equally there are many scholars who claim that postmodern thought is needed more than ever today, to understand social and political constructions of our everchanging world. For example, Holmqvist (2017) believes that postmodern approaches, and in particular earlier writings by Bauman (1992), are highly relevant today as they highlight "how postmodernity is simultaneously a reaction to a previous era and a movement that, again, depends on the one it rebels against for its existence" (p. 150). We agree with Holmqvist that the theoretical foundations of many postmodern theorists are relevant today and can be used to examine the social and institutional construction of a variety of settings and contexts.

Concluding Comments

We conclude this chapter by summarizing the key ideas of postmodern thought to provide a working foundation for the reader to continue with the rest of the book. Throughout the book we return to these ideas and build upon them as well as provide examples to understand how to apply postmodern and poststructural approaches to practice.

- Postmodern thought rejects the concept of an ever onward and upward progress as epitomized by the project of modernity. It disavows the idea that human experience can be reduced to and captured by grand or totalizing theories; the metanarratives of which Lyotard was so critical.
- Rather, postmodern thought emphasizes that reality is plural and that there are multiple positions from which it is possible to view any aspect of reality. This applies to the often taken-for-granted reality of every-day practices. Any writing/speaking/research done about those prac-tices is an attempt to explore, analyse and understand aspects of that reality. It will, and can only ever be a partial analysis, in that it draws on only one of a number of possible positions from which to view the reality in question.
- This has implications for research and researchers. If reality is made up of multiple voices, and if there are multiple positions from which to view that reality, then it follows that no single representation of every-day practices can hope to capture the "truth" about that care or prac-tice. We need to be open to different, shifting, changing and contextual meanings and this focus begins with the researcher locating themselves and declaring a methodological perspective.
- Researchers working within the frame of postmodern thought need to be aware of the role that such thought plays in framing the parameters of the study, the methods employed to obtain 'data,' the analyses that are then done and the conclusions that are reached. What (or who) is absent, or not stated, in any research undertaken is of as much impor-tance as what (or who) is present or stated.
- It is important to remember that postmodern thought is a way of thinking about reality just as any theoretical perspective is. Indeed, one of postmodern's great contributions is to highlight how theory itself, along with the research methods and approaches associated or congru-ent with any particular theoretical orientation, frames our understand-ings of what is appropriate subject matter to study in the first place.

- The unsettling effect of postmodern thought on what we may have come to take for granted in various practices is one of its greatest contributions as it offers possibilities for bringing about change and allowing 'other' voices and perspectives to surface.
- It is important to be aware of the critiques and criticisms of postmodern thought and take up an informed position in relation to them.
- The theoretical foundations of many postmodern theorists are relevant today and can be used to examine the social and institutional constructions of a variety of practice settings and contexts.

Having established a working understanding of postmodern thought that underpins analyses using postmodern approaches, in the next chapter we will take a closer look at the work of one of the most influential theorists associated with postmodern theory – Michel Foucault.

References

Agger, B. (1992). *Cultural Studies as Critical Theory*. London, Falmer Press.

Aylesworth, G. (2015). Postmodernism. In E.N. Zalta (Ed.), *The Stanford Encyclopedia of Philosophy* (Spring 2015 ed.). Metaphysics Research Lab, Stanford University. https://plato.stanford.edu/entries/postmodernism/

Bauman, Z. (1992). *Intimations of Postmodernity*. London, Routledge.

Bauman, Z. (2012). *Liquid Modernity*. Polity Press.

Baxter, J. (2016). Positioning language and identity poststructuralist perspectives. In S. Preece (Ed.), *The Routledge Handbook of Language and Identity*. Routledge.

Bertens, H. (2003). *The Idea of the Postmodern. A History*. London, Routledge, Taylor & Francis Group.

Bertens, H. (2012). Postmodern humanism. *Canadian Review of Comparative Literature*, *39*(3), 299.

Best, S., & Kellner, D. (1991). *Postmodern Theory: Critical Interrogations*. New York, Guilford Press.

Cheek, J. (2008). Healthism: A new conservatism? *Qualitative Health Research*, *18*(7), 974–982.

Cheek, J., & Øby, E. (2023). *Research Design: Why Thinking about Design Matters*. Thousand Oaks, California: SAGE Publications, Inc..

Crawford, R. (1980). Healthism and the medicalization of everyday life. *International Journal of Health Services*, *10*(3), 365–388.

Daniel, S. (1995). Postmodernity, Poststructuralism, and the Historiography of Modem Philosophy, *International Philosophical Quarterly, XXXV*, *3*(139), 255–267.

Eagleton, T. (1983). *Literary Theory: An Introduction*. Minneapolis, University of Minnesota Press.

Finlay, L., & Gough, B. (2003). Introducing reflexivity. In *Reflexivity: A Practical Guide for Researchers in Health and Social Sciences*, Oxford, Blackwell Science Ltd., pp. 1–2.

Foucault, M. (1988). Practicing criticism. In D.R. Rothwell and D.L. VanderZwaag (Eds.), *Politics, Philosophy, Culture and Other Writings*. London and New York, Routledge, Taylor, & Francis Group, 1977–1984.

Holmqvist, A.K.J. (2017). Modernity and Postmodernity in Zygmunt Bauman's Thoughts. *Epokhe Journal of Social Science*, *1*(1), 145–153.

Holtzhausen, D.R. (2013). *Public Relations as Activism: Postmodern Approaches to Theory & Practice*. Routledge.

Kuntz, M. (2012). The postmodern assault on science. *EMBO Reports*, *13*(10), 885–889. https://doi.org/10.1038/embor.2012.130

Lumsden, K. (2019). *Reflexivity: Theory, Method, and Practice*. London, Routledge.

Lyotard, J. (1984). *The Postmodern Condition: A Report on Knowledge*. Minneapolis, University of Minnesota Press.

Mease, J.J. (2017). Postmodern/poststructural approaches. In *The International Encyclopedia of Organizational Communication*. Craig R. Scott & Laurie Lewis (Editors-in-Chief), James R. Barker, Joann Keyton, Timothy Kuhn, & Paaige K. Turner (Associate Editors). John Wiley & Sons. https://doi.org/10.1002/9781118955567.wbieoc167

Murali, K.P. (2019). End of life decision-making: Watson's theory of human caring. *Nursing Science Quarterly*, *33*(1). https://doi.org/10.1177/0894318419881807

Niles, R. (2010). Postmodernism – Postmodern Worldview. https://www.youtube.com/watch?v=ynMS0g6zu5E&t=20s Retrieved July 18, 2023.

Parton, N. (1994). The nature of social work under the conditions of postmodernity. *Social Work and Social Sciences Review*, *5*(2), 93–112.

Richardson, L. (1994). Writing: A Method of Inquiry. In N. Denzin & Y. Lincoln (Eds.), *Handbook of Qualitative Research* Sage, California, pp. 516–529.

Sarup, M. (1989). *An Introductory Guide to Post-structuralism and Postmodernism*. Athens, The University of Georgia Press.

Scheurich, J. (2014). *Research Method in the Postmodern*. London, Routledge.

Susen, S. (2015). *The 'Postmodern Turn' in the Social Sciences*. London and New York, Palgrave Macmillan.

Vattimo, G. (2007). *Nihilism & Emancipation. Ethics, Politics, & Law*. New York, Columbia University Press.

3 The Work of Michel Foucault

In this chapter we will

- Consider the work of Michel Foucault
- Introduce the concepts of knowledge and power
- Discuss the concept of discourse and discursive frameworks
- Examine Foucault's thoughts on governmentality, the clinic, and the panoptic gaze
- Present critiques of Foucault's work
- Provide examples of research that use postmodern concepts drawing on the work of Foucault

Introduction

One theorist whose work has been consistently associated with postmodern perspectives is the French social theorist Michel Foucault. However, Foucault himself resisted the placement of his work in any particular theoretical category, describing it instead as a "history of the present" (Foucault, 1977, p. 31). Further, Foucault's work is very difficult to categorize in a traditional disciplinary sense. As Foster (1985) asks, "is the work of Michel Foucault... to be called philosophy, history, social theory or political science?" (p. x). Such difficulty stems, at least in part, from the fact that Foucauldian analyses, in keeping with postmodern analyses, are "'in between' and 'across' established boundaries of knowledge" (Dean, 1994, p. 13).

Despite such uncertainty in explicitly labelling or categorizing Foucault's work, an exploration of aspects of his work will illuminate some of the understandings of postmodern approaches posited previously in Chapter 2. It will also set the stage for the discussion of post structural thinking in Chapter 4.

DOI: 10.4324/9780429053764-3

Discourse

We will begin the exploration of Foucault's work by focusing on his problematization of knowledge. In keeping with postmodern approaches, Foucault challenges notions which hold that knowledge is objective and value-free, inevitably progressive, and universal. He argues instead that knowledge is inextricably bound to power. Foucault explores the knowledge/power link through the concept of discourse. For Foucault "discourse" refers to ways of thinking and speaking about aspects of reality:

> A discourse provides a set of possible statements about a given area, and organizes and gives structure to the manner in which a particular topic, object, process is to be talked about.
>
> (Kress, 1985, p. 7)

Discourses create discursive frameworks which order reality in a certain way. They both enable and constrain the production of knowledge in that they allow for certain ways of thinking about reality whilst excluding others. In this way they determine who can speak, when, and with what authority, and conversely, who cannot (Ball, 1990). In analysing the effect of such discursive frames, Foucault asks, "what rules permit certain statements to be made; what rules order these statements; what rules permit us to identify some statements as true and some false; what rules allow for construction of a map, model, or classificatory system?" (Philp, 1985, p. 69).

It is important to recognize that at any point in time there are a number of possible discursive frames for thinking, writing and speaking about aspects of reality. However, not all discourses are afforded equal presence or, therefore, equal authority. At any time in history certain discourses will operate in such a way as to marginalize or even exclude others.

Which discursive frame is afforded presence at any time is a consequence of the effect of power relations. "Discourses represent political interests and in consequence are constantly vying for status or power" (Weedon, 1987, p. 41). Indeed Foucault (1984) declares: "Discourse is the power which is to be seized" (p. 110).

Thus, in Foucault's analysis, power is a productive concept: it is not simply repressive. It is the operation of webs of power that enables certain knowledge to be produced and "known."

Paradoxically, such power also constrains what it is possible to know in certain situations.

Power as Capillary and Productive

Foucault was particularly interested in the operation of power at various levels of society, including the level of the health care clinic. He conceived power as being capillary: operating in all levels and directions of society in an extensive network of power relations. Such a view rejects the notion of power as emanating from the top. It also brings into question the notion that society can be divided into simplistic dichotomies, such as the dichotomy of those who have power and those who do not. As Foucault (1980) himself puts it:

> one should not assume a massive and primal condition of domination, a binary structure with "dominators" on one side and "dominated" on the other, but rather a multiform production of relations of domination.
>
> (p. 142)

Foucault was concerned with the effects of power at the very ends of the capillary network, that is, at the sites of its actions such as practice areas like the school, or the health care clinic "rather than at some conjectured sovereign point (the state, the law, or wherever)" (Fox, 1993a, p. 31).

Knowledge from within one discourse can be used to *exclude* knowledge from other discourses. The fact that some discourses gain prominence over others is the result of socio-historical influences operating upon them (Cheek & Rudge, 1994). They achieve "truth" status where truth "is an effect of the rules of a discourse" (Cheek & Rudge, 1993, p. 275).

For example, historically in health care the truth status of medical/scientific discursive frames has shaped dominant taken-for-granted understandings of what is appropriate and authoritative practice. More than 30 years ago Turner (1987) wrote:

> The power of the [medical] profession... depends at least in part, on the ability to make claims successfully about the scientific value of their work and the way in which their professional knowledge is grounded in precise, accurate and reliable scientific information.
>
> (p. 217)

The ability and right of certain groups of health professions rather than others to speak authoritatively about health and illness is premised on the authority of the scientific/medical discourse from which their expertise is both derived and in turn legitimated. This power also depends on the ability to exclude or marginalize other ways of thinking about health care and health care practice, often relegating these other knowledges to the realm of "alternative" health care practices, rather than the mainframe of authoritative contemporary health care.

For example, Young et al. (2019) reported findings from their study that examined how women with endometriosis negotiated power and knowledge during interactions with doctors. Guided by a theoretical frame shaped by feminist social constructionist perspectives, and concepts put forth by Foucault, their analysis concluded among many things that "Women were wary of the social status and power of doctors to reduce their wellbeing through medical labels they did not identify with or by inhibiting their access to care" (Para 1).

Once such exclusionary or marginalizing practices are exposed, it may be possible to interrogate them, to explore which practices seem to dominate, and which are relegated to the margins. Research questions that arise from such analysis include how such a discursive position is maintained, who has an interest in such a maintenance and what the effect is of such discursive practices on practices in the health care sector. This can add a multi-layered, multi-dimensional perspective to the aspect of health care in question. It is not a case of attempting to replace one discourse with another or of using one discourse to exclude others. We explore this point further in the Box below using nursing and nursing practice as an example.

How nurses and nursing practice are portrayed, both by themselves and by others, is to a large extent the result of powers and practices that operate to both construct and position nursing in one way rather than another. The aspects of nursing that are given prominence and the aspects that are not, or are absent, exposes the dominant discursive frames shaping understandings of nursing and nurses at any point in time. It is these discursive frames "and their associated norms and values, which nurses then carry with them into their everyday roles" (Gilbert, 1995, p. 870).

For example, during the COVID-19 pandemic, nurses were predominantly portrayed through the media and around the globe as heroes (Mohammed et al., 2021). This discourse was created to present nurses as caring frontline workers who were needed to save lives. While nurses are caring and do save lives, the timely construction and presentation of this particular caring discourse was to offer a selective meaning of nursing and encourage society to perpetuate these beliefs to try and keep nurses in practice by stereotyping and normalizing their work as altruistic and saving patients. However, nurses pushed back to counter this discourse that was oppressive and harmful to nurses. They did so by using a different discourse that showed the gruelling work and even death of nurses. For example, teenVOGUE begins one of their articles with a sign "Please don't call me a hero, I'm being martyred against my will" (Einboden, 2020, p. 345).

Thus, "the task is not to look for *real* and authentic representations of nursing, but rather to look for the speaking and representation that is done about nursing" (Cheek, 1995, p. 239) and who

> gains and who loses from such representation and speaking. In keeping with this, Gilbert (1995) asserts that:
>
>> The nurse needs to be able to identify the discursive practices through which they as nurses are formed. For it is these, and their associated norms and values, which nurses then carry with them into their everyday roles.
>>
>> (p. 870)

The Clinic as a Discursive Construction

Foucault was particularly interested in the emergence of the clinic as central to health care practice. The birth of the clinic was possible because of a disease-based model of health care. Foucault (1975) asserted that the clinic was the institutionalization of the medical gaze: "a hearing gaze and a speaking gaze: clinical experience represents a moment of balance between speech and spectacle" (p. 115). By *gaze*, Foucault was referring to the act of seeing, or the way in which disease, illness and health care were thought about and viewed.

The act of examining the body is a central tenet of the construction of health care around the locus of the clinic. In the examination, the body is made the object of the health professional's gaze. The body is scrutinized within the parameters of the scientific/medical discursive frame. The objectified body is then subjected to the regimes of truth and to the technologies of power created by the understandings of this discursive frame in the form of tests, procedures, and further examinations. Examination of the body in this way allows for the development of further scientific/medical knowledge.

The body is thus productive of certain forms of knowledge and in turn, subjugated to the discipline of these knowledges (Foucault, 1977). Foucault (1975) suggests that the resultant disciplined body becomes a docile body which is the subject and object of the clinical gaze and its underlying premise of the legitimacy and authority of scientific/medical discourse.

In Foucault's analysis, the emergence of the clinic itself was only possible because of the emergence of certain types of discursive frames and associated knowledges. The clinic, in turn, operated to further produce certain types of knowledge whilst excluding others. Hence Long (1992) describes the clinic as:

> a mode of perception and enunciation that enables statements about disease. The clinic is a discourse because it is a set of rules and procedures within the field of inquiry... the clinic is more than a place where medicine is practiced, it is also a discursive practice, a language of health and disease.
>
> (pp. 119–120)

Or put another way, the clinic is both produced by, and in turn produces, dominant discourses related to health and health care practices. Dominant discourses such as what health and health care practices are/ are not; and how, where and by whom they might be enacted/not enacted.

Governmentality

This close relationship between power and knowledge can be seen in the connections between modern forms of governance and the discourses of human sciences such as medicine, psychiatry, and law. Such government "implies all those tactics, strategies, techniques, programmes, dreams and aspirations of those authorities that shape beliefs and the conduct of the population" (Nettleton, 1991, p. 99). According to Foucault (1979), governmentality is characterized by pervasive matrices of power which entail the surveillance and disciplining of both individuals and entire populations: the population is both subject and object of government. There is a network of power in that "the State is superstructural in relation to a whole series of power networks that invest in the body, sexuality, the family, kinship, knowledge, technology and so forth" (Foucault, 1980, p. 122). Governmentality reinforces the "community" whilst at the same time (somewhat paradoxically), increasing individualization. Fox (1993a) defines governmentality as the "subtle, comprehensive management of life drawing both from a top-down exercise of power over conduct...with a subjectivity constituted in a sense of personal responsibility, rights, freedoms and dependencies" (pp. 32–33).

In such a conception, power is diffuse, anonymous, and subtle, with its associated disciplinary regimes and techniques producing individuality. The individual is the site of the operation of powerful discourses and is in effect a product of the inscription of such discourses. Thus:

> the individual is both the site for a range of possible forms of subjectivity and, at any particular moment of thought or speech, a subject, subjected to the regime of meaning of a particular discourse and enabled to act accordingly.
>
> (Weedon, 1987, p. 34)

At the same time, registers of births and deaths and reports of certain diseases and other health-related statistics enable the monitoring of trends in disease and illness in entire populations. These trends can then be used to establish the norm and to further regulate and discipline the behaviour of both individuals and entire populations, subjecting them increasingly to the gaze of the health professional's authority.

According to Foucault (1982), "to govern, in this sense, is to structure the possible field of action of others" (p. 221). By its very nature, the exercise of this power relies on the knowledge of experts, for it is they who decide what is normal and abnormal within populations, and it is they who identify abnormality in individuals.

"Authorities of various sorts have sought to shape, normalize and instrumentalize the conduct, thought, decisions and aspirations of others in order to achieve the objectives they consider desirable" (Miller & Rose, 1990, p. 8). The human body is thus a political body as well as corporeal, and it is "this political body which is the locus for the politics of health-talk, not the anatomical body which medicine and its adjunct disciplines have fabricated" (Fox, 1993a, p. 24).

Foucault's analysis highlights the ongoing quest to normalize understandings and functions of the human body. There are "normal" functions of the body able to be delineated within normal understandings of appropriate ranges and limits. There are "normal" ways of practicing health care and treating illness/disease. Individuals are exhorted to attain, for example, "normal" ranges of height, weight, and behaviour. Disciplinary regimes are instituted for those who do not measure up according to the health gaze. Such disciplinary regimes involve individuals being under the constant scrutiny or gaze of a health "expert," the individuals themselves, or both the expert and individuals. Accompanying such scrutiny are disciplinary regimes such as diet, exercise or other more technical procedures all designed to restore the individual to the norm (see the discussion of healthism in Chapter 2 which provides examples of this) The bodies of individuals are thus made visible and inscribed by the discourse of health professionals (Fox, 1993a).

The capacity of experts in the human sciences to perform their functions of surveillance, categorization and intervention presupposes a set of power relations which enables them to carry out such activities. More profoundly, their discourses—the fields of possibility of their knowledge—about what counts as sickness or health, madness or sanity, criminality or lawfulness are the results of power relations. Medicine, with its knowledge of human bodies and its capacity to survey them, is an important mechanism of the modern disciplinary regime. The power of scientific/medical frames in contemporary health care is evident in that:

> Not only can these experts define the objects of their study, such as for example what constitutes a sick or insane client, but they also determine the limit of possibilities for the study, treatment and management of the objectified client. The power and control exercised by "experts" over clients-as-objects thereby constrain the types of discourse that are able to be developed about the client's condition.
>
> (Cheek & Rudge, 1993, p. 276)

The Idea of the Panoptic Gaze

Foucault uses the metaphor of the panopticon to encapsulate his argument about the surveillance of individuals and even of whole populations by the gaze of "experts." The panopticon, as described by Jeremy Bentham (2012) was a circular prison in which all cells were open and faced inwards towards a single central guard tower. All prisoners were thus under the potential scrutiny not only of the guard in the tower but of each other at any particular moment. Similarly, the guard could be scrutinized at any time by the prison governor. Consequently, a network of surveillance was created in which each individual was under the potential examination and gaze of a number of individuals at any one time. Not knowing when one was actually under scrutiny, or by whom, resulted in self-disciplinary, compliant and docile behaviour because individuals acted as though they were under scrutiny at all times.

Foucault posited that such panoptic tendencies were evident in contemporary health care: in the examination of individuals by health professionals, in the individuals' self-policing of their own health, in the control of whole populations in the name of public health, and in the quest to normalize such populations. Individuals and populations could also be scrutinized by others and by themselves in regard to risk behaviours, lifestyle, and susceptibility to certain diseases and illnesses. They stand perpetually before the gaze of a panoptical health connoisseur (Bartky, 1988), who is able to scrutinize all aspects of their body and health. Foucault suggests that a consequence of such scrutiny is the development of docile bodies complicit in their own scrutiny and policing.

According to Foucauldian analysis, the examination of individuals by the gaze of experts such as those in the health care field is a disciplinary technique of power that invents "a new kind of individuality" (Hoskin, 1990, p. 33), namely "the individuality…of the calculable man *[sic]*" (Foucault, 1977, p. 193). Health practice is premised on confining or restoring individuals to the norm—a calculated and calculable norm beyond which anything or anyone is "other." In Foucauldian analysis, modern scientific/ medical discourse is only one of many discursive frames within which health and illness might be conceptualized. The dominance of this discourse lies not in its rationality or logic but in the power that both underpins and maintains its discourse. It lies in the ability of its proponents to exclude others from practising different forms of health care and in its ability to relegate other ways of thinking about health care to just that— the "other."

This is why Foucault (1984) states that:

> discourse is not simply that which translates struggles or systems of domination, but is the thing for which and by which there is struggle, discourse is the power which is to be seized.

> (p. 110)

Standing before the gaze

Aston et al. (2016) conducted research on mothers' and public health nurses' experiences during postpartum home visits. They found that there was a dominant medical and social breastfeeding discourse that 'expected mothers to breastfeed.' This created a type of self-surveillance and affected the way mothers felt about themselves when choosing a feeding method for their babies. A dominant discourse of 'breast is best' often made mothers feel guilty if they could not, or chose not to breastfeed. The study concluded that mothers had to continually negotiate relations of power related to dominant medical and social discourses on child rearing. You can read the full study at:

Aston, M., Etowa, J., Price, S., Vukic, A., Hart, C., MacLeod, E., Randel, P. (2016). Public health nurses and mothers challenge and shift the meaning of health outcomes. *Global Qualitative Nursing Research* 3, 1–10. doi:10.1177/2333393616632126

Using Foucauldian Thought in Research About Practice

The clinic, the "ward round," and the "consultation" have emerged as normalized, discursively constructed organizational strategies underpinning contemporary health and health care understandings and practices. The consultation and the ward round and all that they entail in terms of setting, professional discourse, and so forth are not the only ways in which the interaction between patients and health practitioners might be organized. Yet, they have become a taken-for-granted aspect of the organization of health care. Postmodern research approaches give the opportunity to ask questions such as why these organizational strategies exist, how and why they are maintained, how they affect understandings of health and health care, whose interest they serve, and how they might be otherwise. This is true of other practice areas.

So, how can such a theoretical frame such as postmodern thought be used in research? What might a postmodern approach to research about a particular focus or area in health care look like? A good way to explore these questions is to look at ways in which researchers have drawn on postmodern approaches to inform their investigation of a particular practice or applied area. For example, check out how:

1   Nettleton (1991) explored the importance of the gaze in contemporary health care in her study of governmentality in dental care. She

demonstrated the network of power surrounding the development of dental hygiene techniques and procedures. Nettleton was particularly interested in exploring how the advent of the dental examination led to an aggregation of data on the "mouths of the population" (p. 101). This resulted in claims by dentists to expert knowledge and "wisdom" about the management and surveillance of dental health care. Nettleton argues that concomitant with the development of such "wisdom" we see the rise of other agents, such as hygienists, health visitors and parents to assist in restoring and maintaining that norm. Consequently, webs of governmentality are developed concerning dental hygiene, techniques, and procedures. So too are "experts" situated at various sites such as the school, the dental surgery, and the home.

Nettleton focused on the dynamic between the exercise of power at the macro level (such as in public health screening, establishing population norms for dental care and so forth) and the individual's complicity with such a gaze by adopting regimes of self-surveillance. In light of such a gaze and being scrutinized;

> There is no need for arms, physical violence, material constraints. Just a gaze. An inspecting gaze.... each individual thus exercising this surveillance over, and against, himself [sic]. A superb formula: power exercised continuously and for what turns out to be a minimal cost.
>
> (Foucault, 1980, p. 155)

Her analysis highlights how Foucauldian and postmodern analyses can illuminate and challenge aspects of the everyday reality of health care, many of which have become so taken for granted that they are assumed to be "normal" or the way things "should be."

2 Fox (1993b, 1994), used ethnographic techniques and interviews, to explore both the surgical operation and the surgical ward-round (sometimes referred to as "grand rounds") as organizational strategies that were constituted by the discourses of the various professionals who worked within those strategies. He did not analyse the ward-round in terms of how, when and by whom it was conducted. Rather, as with all discursive analysis, he looked critically at the idea and practices of the ward-round as an organizational strategy *itself*. Questions which arose from such analysis included: what enabled an organizational strategy such as a ward-round to be established in the first place, and who or what subsequently enabled it to be maintained in that form? What part did the discursively framed understandings of the various health professionals involved in the ward-round, particularly the surgeons themselves, play in constructing the round? Conversely, how did the ward-round itself, and the discursive assumptions embedded within it, in turn frame understandings about these various health professionals?

Finally, what about the "patient"? How did the way the ward-round's constituted position, shape understandings about the patient in relation to the other health professionals? Similarly, May et al. (1996) explored the consultation between doctors and patients in general practice as an organizational strategy that frames the relationship between doctor and patient. May et al. focus on the ways in which the professional rhetoric of general practice has come to be organized around a very specific view of the possibilities that arise from this relationship. The consultation, like the ward-round, is shaped discursively and in turn shapes the way the general practitioner and the patient are viewed by each other, by themselves and by others.

3 Peckover and Aston (2017) discussed the concept of surveillance as described by Foucault and applied it to analysis of surveillance and governmentality of mothers with young children by public health nurses (PHNs) or home visitors (HVs). While the intent of the nurses was to support mothers in the home during the postpartum period, nurses were also viewed as monitoring or surveying the practices of mothers who sometimes feared they would remove their babies due to perceived inadequate parenting practices. They wrote:

> Conclusions and Relevance to Clinical Practice: In this paper we have critically examined the HVs and PHNs surveillance literature that demonstrates how the embedded practices of surveillance by HVs and PHNs are socially and institutionally constructed through complex diverse binary and overlapping discourses. Applying a Foucauldian lens enables one to see how relations of power create particular interactions between mothers and HVs or PHNs. These interactions guide the type of support and care that mothers receive – all framed within a surveillance lens.
>
> (p. 22)

4 Berg (2020) completed an autoethnographic study critically examining her experience as a dance teacher and observations with adolescent female dancers. Using "Foucauldian theory as well as surveillance and social media scholarship" (p. 135) Berg discussed the impact of live-streamed closed-circuit television and the imagined audience for whom student dancers performed. Berg showed how dance teachers used their power to regulate dance behaviours with dancers by simply observing or gazing with very little spoken words. Such surveillance maintained normalized understandings of how dance should be practiced including how dancers' bodies should look, act, and move. Such normalized understandings acted to perpetuate gendered stereotypes about female and male bodies. Dancers themselves, administrators, and teachers, became both subject and object of such surveillance, thereby all perpetuated these unspoken norms through self-surveillance and the surveillance of others.

These examples highlight how postmodern analysis drawing on Foucauldian thought offers the possibility for new and/or different discourses to surface in practice areas. It enables the potential for the multidimensional and multi-perspective nature of practice to be represented through the frames of multiple discourses. A postmodern perspective allows for the analysis of why practices have been shaped in the way they are, and why certain players and practices in areas such as health care delivery and dance education have been relegated to the margins, often designated as "other" rather than "another." This type of analysis may well form the basis of a postmodern-oriented research endeavour.

It is important to note that Foucault resisted giving "a" method for doing Foucauldian-oriented research:

> Despite a concern with discourses as rule-governed systems for the production of thought, Foucault never sought to apply a particular system or to allow his own heuristics to congeal into a fixed, formal method.
>
> (Dean, 1994, p. 14)

To give "a" method would be to create another dominant discourse that could then be used to exclude other approaches – the antithesis of Foucauldian inspired thought.

Critiques of Foucault's Work

Foucault's work has not been without its critics. Collins (1990) scathingly referred to Foucault as a theoretical "amateur" (p. 462) in that his theory was not developed or presented within conventional theoretical frames. However, as Agger (1992) points out:

> Collins does not recognize that Foucault would have loved to be called an amateur. Foucault implies that the professional/amateur distinction is a peculiar product of the discourse/practice of late capitalism, wherein unofficial knowledges are disqualified as unrigorous, undisciplined, unprofessional.
>
> (p. 126)

Further, like postmodern thought more generally, Foucauldian analyses are often criticized as being nihilistic – pulling everything down and leaving little in their wake. As Waltzer (1988) puts it, "angrily he rattles the bars of the iron cage. But he has no plans or projects for turning the cage into something more like a human home" (p. 209). In relation to the health care context, Porter (Cheek & Porter, 1997) echoes Waltzer's critique when he writes "while Foucauldian analysis can tell us a lot about what is wrong with where we are, it can tell us very little about where we should go" (p. 113).

However, it could be argued that Foucault's major contribution is to expose the bars of what previously may have been a largely invisible, and thus taken-for-granted, iron cage. As Hoy (1991a) points out (drawing on the same analogy of the prisoner used by Waltzer):

> Foucault's voice may sound like that of the prisoner who wants out and cannot get out, but since Foucault is talking about our inability to get out of our own place in history, he is surely correct in this regard.
>
> (pp. 143–44)

Foucault never claimed to offer a total picture with all the answers. As Hoy (1991b) points out, Foucault rejected the "traditional philosophical goal of constructing a total theory that can explain the entire social reality" (p. 5). Foucault does not offer the possibility of a power-less situation either in the health context or elsewhere. As Smart (1991) asserts:

> It is clear therefore that in Foucault's terms there can be no power-free or power-less society, no millennial end of history towards which oppressed, exploited or dominated subjects may be led or guided, for relations of power, that is, ways of acting upon the actions of (other) acting subjects, are endemic in society.
>
> (pp. 169–70)

What Foucault does offer us, however, is a way of *understanding* power and its effects, rather than a grand vision of how power might be overcome. Instead of working to overcome or eliminate power entirely, it may be possible to work with and or against different sites of the capillary relations of power that pervade any context, including the health care setting.

We are beginning to move into some complex arguments here that are beyond the scope of an introductory text about postmodern and poststructural approaches. However, this brief and limited discussion has pointed to the complexity of Foucault's work. A fuller discussion of the possibilities and problems posed by the use of Foucauldian theory in nursing and health care is provided in an article "Reviewing Foucault: Possibilities and problems for nursing and health care" which Julianne co-authored with Sam Porter. (Cheek & Porter, 1997). While written some decades ago the discussion remains just as relevant today.

In this paper, which is written in the form of a dialogue between Sam Porter and Julianne on the relative merit of the contribution of

Foucauldian theory to nursing and health care, many of the points briefly alluded to previously in this chapter are developed more fully. If you are interested in reading more on Foucault this article makes an ideal starting point. The article concludes by cautioning against "bad short answers" (Hacking, 1991, p. 27) to complex theoretical questions. Likewise, we wish to conclude this introduction to Foucault's work by reiterating that the discussion has necessarily been a brief overview. However, as is the case with our dialogue about the potential and problems of Foucault's work, "while the analysis offered here has indeed been modest in scope, the implications of what has been proposed are anything but" (Cheek & Porter, 1997, p. 117).

Foucault: Some Concluding Remarks

Foucault's work provides a useful way of exploring what postmodern approaches can tell us about aspects of contemporary health care and health care practice. His concept of discourse-that is, certain ways of thinking and talking about reality—demonstrates the inextricable link between power and knowledge. Analysis of the impact of such discursive understandings opens up entirely new fields and foci for research in practice areas such as health care and education.

Foucault's concept of the panoptic gaze demonstrates the centrality of the examination as an outworking of the gaze in contemporary practice settings, and the production of docile individuals as objects of, and as subject to, such a gaze.

In the latter part of the chapter, examples of research that used Foucauldian concepts were discussed in order to demonstrate how Foucauldian thought and research can be operationalized in the world of practice and thereby provide understandings of that practice world.

The next chapter considers poststructural thought and its role in informing research related to practice and practice areas.

References

Agger, B. (1992). *Cultural Studies as Critical Theory*. London, Falmer Press.

Aston, M., Etowa, J., Price, S., Vukic, A., Hart, C., MacLeod, E., & Randel, P. (2016). Public health nurses and mothers challenge and shift the meaning of health outcomes. *Global Qualitative Nursing Research*, *3*, 1–10. https://doi.org/10.1177/2333393616632126

Ball, S. (1990). Introducing Monsieur Foucault. In S.J. Ball (Ed.), *Foucault and Education: Disciplines and Knowledge*. London, Routledge, pp. 1–8.

Bartky, S. (1988). Foucault, femininity, and the modernization of patriarchal power. In I. Diamond & L. Quinby (Eds.), *Feminism and Foucault: Reflections on Resistance*. Boston, Northeastern University Press, pp. 61–68.

Bentham, J. (2012). The panopticon. In P. Priestley & M. Vanstone (Eds.), *Offenders or Citizens?* London and New York, Routledge, Taylor & Francis Group, pp. 13–15. Willan.

Berg, T. (2020). Manifestations of surveillance in private sector dance education: The implicit challenges of integrating technology. *Research in Dance Education*, 135–152. https://doi.org/10.1080/14647893.2020.1798393

Cheek, J. (1995). Nurses, nursing and representation: An exploration of the effect of viewing positions on the textual portrayal of nursing. *Nursing Inquiry*, *2*, 235–240.

Cheek, J., & Porter, S. (1997). Reviewing Foucault: Possibilities and problems for nursing and health care. *Nursing Inquiry*, *4*, 108–119.

Cheek, J., & Rudge, T. (1993). The power of normalisation: Foucauldian perspectives on contemporary health care practices. *Australian Journal of Social Issues*, *28*(4), 271–284.

Cheek, J., & Rudge, T. (1994). Nursing as textually mediated reality. *Nursing Inquiry*, *1*(1), 15–22.

Collins, R. (1990). Cumulation and anti-cumulation in sociology, *American Sociological Review*, *55*, 462–463.

Dean, M. (1994). *Critical and Effective Histories: Foucault's Methods and Historical Sociology*. London, Routledge.

Einboden, R. (2020). SuperNurse? Troubling the hero discourse in COVID times. *Health*, *24*(4), 343–347. https://doi.org/10.1177/1363459320934280

Foster, H. (1985). Postmodernism: A preface. In *Postmodern Culture*. London, Pluto Press, pp. ix–xvi.

Foucault, M. (1975). *The Birth of the Clinic*. New York, Vintage Books.

Foucault, M. (1977). *Discipline and Punish*. London, Tavistock.

Foucault, M (1979). Governmentality. *I & C*, *5*, 5–21.

Foucault, M. (1980). *Power/Knowledge*, C. Gordon (Ed.) Brighton, Harvester Press.

Foucault, M. (1982). Afterword: The subject and power. In H. Dreyfus & P. Rabinow (Eds.), *Michel Foucault: Beyond Structuralism and Hermeneutics*. Chicago, University of Chicago Press, pp. 208–226.

Foucault, M. (1984). The order of discourse. In M. Shapiro (Ed.), *Language and Politics*. Oxford, Basil Blackwell, pp. 108–138.

Fox, N. (1993a). *Postmodernism, Sociology and Health*. Toronto, University of Toronto Press.

Fox, N. (1993b). Discourse, organisation and the Surgical Ward Round. *Sociology of Health and Illness*, *15*(4), 16–42.

Fox, N. (1994). Anaesthetists, the discourse on patient fitness and the organisation of surgery. *Sociology of Health and Illness*, *16*(1), 1–18.

Gilbert, T. (1995). Nursing: Empowerment and the problem of power. *Journal of Advanced Nursing*, *21*, 865–871.

Hacking, I. (1991). The archaeology of Foucault. In D. Hoy (Ed.), *Foucault: A Critical Reader*. Oxford, Basil Blackwell, pp. 27–40.

Hoskin, K. (1990). Foucault under examination: The crypto-educationalist unmasked. In S. Ball (Ed.), *Foucault and Education: Disciplines and Knowledge*. London, Routledge, pp. 29–53.

Hoy, D. (1991a). Power, repression, progress: Foucault, Lukes and the Frankfurt School. In D. Hoy (Ed.), *Foucault: A Critical Reader*. Oxford, Basil Blackwell, pp. 123–148.

Hoy, D. (1991b). Introduction. In D. Hoy (Ed.), *Foucault: A Critical Reader*. Oxford, Basil Blackwell, pp. 1–26.

Kress, G. (1985). *Linguistic Processes in Socio-cultural Practice*. Victoria, Deakin University Press.

Long, J. (1992). Foucault's clinic. *Journal of Medical Humanities*, *13*(3), 119–138.

May, C., Dowrick, C., & Richardson, M. (1996). The confidential patient: The social construction of therapeutic relationships in general medical practice. *Sociological Review*, *44*(2), 187–203.

Miller, P., & Rose, N. (1990). Governing economic life. *Economy and Society*, *19*(1), 1–31.

Mohammed, S., Peter, E., Killackey, T., & Maciver, J. (2021). The "nurse as hero" discourse in the COVID-19 pandemic: A poststructural discourse analysis. *International Journal of Nursing Studies*, *117*, 103887.

Nettleton, S. (1991) Wisdom, diligence and teeth: Discursive practices and the creation of mothers. *Sociology of Health and Illness*, *13*(1), 98–111.

Peckover, S., & Aston, M. (2017). Examining the social construction of surveillance: A critical issue for health visitors and public health nurses working with mothers and children. *Journal of Clinical Nursing*, *27*(1–2), e379–e389. https://doi.org/10.1111/jocn.14014

Philp, M. (1985). Michel Foucault. In Q. Skinner (Ed.), *The Return of Grand Theory in the Human Sciences*. Cambridge University Press, pp. 67–81.

Smart, B. (1991). The politics of truth and the problem of hegemony. In D. Hoy (Ed.), *Foucault: A Critical Reader*. Oxford, Basil Blackwell, pp. 157–174.

Turner, B. (1987). *Medical Power and Social Knowledge*. London, Tavistock.

Waltzer, M. (1988). *The Company of Critics*. New York: Basic Books.

Weedon, C. (1987). *Feminist Practice and Post Structuralist Theory*. London, Basil Blackwell.

Young, K., Fisher, J., & Kirkman, M. (2019). Partners instead of patients: Women negotiating power and knowledge within medical encounters for endometriosis. *Feminism and Psychology*, *30*(1). https://doi.org/10.1177/0959353519826170

4 Thinking and Researching Poststructurally

> **In this chapter we will**
>
> - Introduce the concept of text as a central part of poststructural thought
> - Discuss how different practices are textually mediated
> - Explore the concept of deconstruction
> - Investigate how deconstruction inspired thinking utilizes the idea of binary opposites
> - Demonstrates use and the possibilities of deconstruction for analysing textual representations of practice

The Centrality of the Text in Poststructural Thought

In chapter one we developed a working understanding of poststructural thought that we are using in this book. This was that poststructural perspectives challenge the notion that language is a neutral, objective, value-free conveyer of aspects of reality. They expose and interrogate language itself as being both constituted by, and constitutive of, the social reality that it seeks to represent. Thus, it is the task of poststructural research approaches to "investigate the meaning of particular representations: to understand how they came to be as they are, and what they communicate about their specific cultural and historical contexts" (Squier, 1993, p. 30).

As we noted in the opening chapter, poststructural perspectives have much in common with postmodern perspectives, so much so that some writers have used the terms synonymously. However, poststructural and postmodern perspectives differ in their focus and emphasis. Unlike postmodern analyses which tend to be wider in scope and which focus on aspects of culture, society, and history, poststructural studies have tended to concentrate on analyses of literary and cultural *texts*, where *text* refers

DOI: 10.4324/9780429053764-4

to a representation of any aspect of reality (Agger, 1991). Mease (2017) sums up the differences between these approaches as being:

> One might describe postmodernism as a particular way of doing or being that challenges the conventionally accepted notion of universal truths and norms by playing with and embracing alternatives to those truths and norms. Poststructuralism offers a more specific academic project that emerged from the study of language, its uses, and the ways it structures lived experience.
>
> (Par 4)

Derrida (1976), an influential and foundational thinker in poststructural textual analysis writes that "nothing is ever outside a text since nothing is ever outside language, and hence incapable of being represented in text" (p. 35). Texts can be pictures, poems, procedures, conversations, case notes, artwork or articles. Texts represent "conventionalized practices... which are available to text producers and interpreters in particular social circumstances" (Fairclough, 1992, p. 194).

Agger (1991) referred to these conventionalized practices as "the assumptions that every text makes in presuming that it will be understood" (p. 112) and goes on to point out that "these assumptions are suppressed, and thus the reader's attention is diverted from them" (p. 112). The way a text represents an aspect of reality, that is, the conventionalized practices and assumptions that underpin the shaping of the text itself, is of as much interest as what the text actually describes. Cheek and Rudge's (1994b) study of case notes in a rehabilitation setting is an example of a study that aimed to expose the conventionalized practices and assumptions (most often undeclared) that underpin and shape the practices of producing those case notes.

Case Notes as Texts Produced By, and Productive Of, the Context in Which They Were Developed

Julianne Cheek and Trudy Rudge analysed the case notes that were developed about two patients in a rehabilitation setting. The case notes were examples of texts that were organized by, and in turn helped to organize and normalize conventionalized practices in the rehabilitation setting and context in which the case notes were developed. Their analysis took as its focus the interaction between the text (the case notes) and the context (the rehabilitation setting in which the text was produced). A basic premise of this study was that nursing and other aspects of health care are textually mediated and that analysis of those texts can provide insights into how such texts constitute and reproduce the socially constructed act of nursing (Cheek & Rudge, 1994a, 1994b).

In keeping with the poststructural theoretical frame underpinning their study, Cheek and Rudge (1994b) were clear that "our concern is not whether these are either a poor or good record of the events, but rather the nature of the reality produced by these texts" (p. 43). Such a focus moves the research beyond the descriptive. The focus is not so much on what is recorded as it is on why it is recorded, and conversely, why other things are not recorded. How did, and do, case notes come to be developed in the form they are, and why and how is such a form maintained? And what is the effect of this on practices and assumptions in that context? We will return to this example in Chapter 5.

Cheek and Rudge's study highlights that in research inspired by poststructural thinking, the way practices and procedures are represented, whether they are written, spoken, or acted, *and the effects of that representation*, become the data and the focus of the analysis rather than, for example, how many times a certain procedure or practice is carried out, and how effectively and efficiently. This is to take a step back as it were in terms of the focus of the research to be undertaken. Rather than accepting procedures or practices (e.g., the way case notes are written and used) as a given, those very procedures and practices *themselves* are made the focus of the research.

What Does All This Mean for Research?

Questions that might be asked by a researcher drawing on poststructural analyses about any view or representation of any aspect of reality include whether this is the only way this aspect of practice can be represented; why this representation is the one accepted as "normal" or "given"; what are alternate ways of representing the same reality; and why these ways are absent and/or marginalized and suppressed. Such questioning exposes language as "an important site of political struggle" (Weedon, 1987, p. 24) and language *itself* as being both constituted by, and constitutive of, the social reality that it seeks to represent.

Research drawing on poststructural perspectives focuses on the way texts are "structured by assumptions within which any speaker must operate in order to be heard as meaningful" (Ball, 1990, p. 3). For example, there are "correct" or accepted ways of acting and thinking about practice and practice areas. How does the way these "correct" or "accepted" practices are organized represent certain views about what constitutes "correct," "normal," "optimal," "acceptable" practice? For example, the assessment of students, the conduct of a ward round, or professional dress. Do these views also enable claims of authority to be made along with the ability to dismiss other viewpoints on the basis of that authority. If so how?

Koro-Ljungberg et al.'s (2015) analysis of the practice of peer review as textually mediated provides a good example of this type of questioning. They analysed the peer review process and the way it generated articles constructed by, and in turn constructive of, understandings of what an acceptable scientific article is in terms of what is written in that article and how it is written.

Using post structural thinking, Koro-Ljungberg et al. (2015) problematized aspects of the peer review process and illustrated that "(A)lthough the peer-review event and process as an institutionalized form appears to be objective and rational…it is fraught with various forms of Foucaultian power relations." (p. 28). The texts analysed were letters and reviews that were exchanged between authors, reviewers, and editors during the peer review process of a manuscript submitted for publication.

They show how these texts (letters and reviews) are formed by assumptions about what good science is, what is and is not acceptable science, who says, and who has the power to make their view of "acceptable" stick. In this way these texts associated with peer review both construct and are constructed by dominant discourses about what research is, and is not, and also how it can, and cannot be, talked about.

Koro-Ljungberg, et al.'s (2015) analysis of the peer review process provides a very good example of the way texts (in this case journal articles, reviews of those articles, and editorial communications about those articles) are structured by assumptions about what makes "good" "science" and how that good science may/may not be written. They demonstrate the way that these assumptions influence decisions about what will and will not be published in peer reviewed journals, and who has the power to make these decisions. And what the effects are of those decisions.

Textual Constructions of Practice Areas and Practices in Them

Similarly, there are "correct," or accepted ways of acting and thinking in practice areas such as health and education. This includes thinking about what constitutes "professional" practice, and acceptable/unacceptable forms of that practice. Such thinking also enables authority to be claimed on the basis of legitimate professional practice and enables other viewpoints of that practice to be de-legitimized as "unprofessional" on the basis of that authority. Once understandings of what is deemed professional (and what is not) are established, those understandings can be used to regulate practices and exclude those practices and practitioners not conforming with those regulations.

The emergence of the profession of physiotherapy as a result of, and response to the London massage related scandals in 1894 (Nicholls &

Cheek, 2006) is an interesting example of how views about what constitutes acceptable/unacceptable forms of practice enables claims of authority and the ability to use those claims to dismiss other viewpoints on the basis of that claimed authority.

Using contemporary documents related to these 1894 scandals such as the paper published by the BMJ in 1894 titled 'Astounding Revelations Concerning Supposed Massage Houses or Pandemonium's of Vice,' Nicholls and Cheek (2006) show how this led to the formation of the Society of Trained Masseuses (STM). The STM "acted to legitimize massage, which had become sullied by its association with prostitution...The founders established a clear practice model for massage which effectively regulated the sensual elements of contact between therapist and patient" (p. 2336). This regulation took the form of set curricular, examinations, and inspection of the practices of members of the Society. "A biomechanical model of physical rehabilitation was adopted to enable masseuses to view the body as a machine rather than a sensual being" (p. 2336).

Such biomechanical discourses legitimized physical rehabilitation and gave license to an emerging physiotherapy profession that then could "touch patients, massage and manipulate them, interact and treat them, whilst at the same time addressing the vexed question of legitimacy" (Nicholls & Cheek, 2006, p. 2345). In this way biomechanical discourses enabled physiotherapists to attain social respectability for themselves and their work. Such discourses also provided ways for deciding what is, and is not, legitimate practice and/or practitioners, and then regulating and enforcing those decisions.

If you want to read more about this study including how poststructural (and postmodern) thought was used to frame and underpin the study theoretically, including the Foucauldian notion of discourse as both constructed by and constructive of aspects of social contexts, check out the full article reporting this study at

Nicholls, D. A., & Cheek, J. (2006). Physiotherapy and the shadow of prostitution: The Society of Trained Masseuses and the massage scandals of 1894. *Social Science & Medicine, 62*(9), 2336–2348.

To Sum Up Where Our Discussion Has Taken Us So Far

Poststructural approaches lend themselves to a distinct textual research focus. The representations of practices and procedures, whether they are written, spoken, or acted, are the data and the focus of the analysis rather than, for example, how many times a certain procedure or practice is carried out, and how effectively and efficiently. This is, as it were, to take a step back in terms of the focus of the research to be undertaken. Rather than accepting the reality of a setting as a given, and therefore the starting point for the research, instead that very reality itself is made the focus of the research.

In the next section of the chapter we take a closer look at deconstruction as one way of putting poststructural inspired thinking into practice when designing and conducting our research.

Deconstruction: An Approach to Analysing Texts Using Poststructural Perspectives

Deconstruction is an approach that is associated with the exploration and interrogation of texts using poststructural perspectives. Poststructural perspectives, and in particular deconstruction, are often associated with the work of the French theorist Jacques Derrida. This is even though he, like Foucault, was careful not to classify his work as belonging to any particular theoretical orientation. Nor did he elaborate "a single deconstructive method" (Agger, 1991, p. 112). Rather, his work, just like the term deconstruction itself, represents a range of approaches each with its own emphases.

Consequently, deconstruction does not represent a unitary concept; it involves a multiplicity of fields and styles:

> To present "deconstruction" as if it were a method, a system or a settled body of ideas would be to falsify its nature and lay oneself open to charges of reductive misunderstanding.
>
> (Norris, 1991, p. 1)

Norris goes on to point out that at times, the term deconstruction tends to suffer from a fairly loose usage to refer to any critique of existing social structures. Nevertheless, despite ambiguity and plurality in the way the term "deconstruction" is used, all deconstructive approaches have the same purpose. Rather than seeking to find "the" meaning within, or of, any text, they seek to challenge the very meanings and the assumptions on which those meanings are founded:

> Methodologically, deconstructionism is directed to the interrogation of texts. It involves the attempt to take apart and expose the underlying meanings, biases, and preconceptions that structure the way a text conceptualizes its relation to what it describes.
>
> (Denzin, 1994, p. 185)

Thus, deconstruction involves a particular way of thinking about texts and a particular way of "reading" them: not to find "the" meaning of that text but to trouble the assumptions underpinning the text. Deconstruction involves looking at the representation of reality in the text as a partial representation (Feldman, 1995) exploring silences and gaps in the text and what they reveal. Thus, deconstruction is "less method than perspective, a kind of interpretative self-consciousness" (Agger, 1992, p. 95).

Agger (1992) has highlighted some of the assumptions implicit within a deconstructive approach and it will be useful to briefly review those assumptions here. The first assumption is that culture is a text and can therefore be interrogated as a text: "the boundary between the textual world and the social world fades once we subject culture to a deconstructive reading" (Agger, 1992, p. 98).

The second assumption is that deconstruction aims to locate the author. By this Agger is alluding to the unwritten and unspoken assumptions that the author has made by selecting what will, and conversely will not, be said or written in the text and by determining what form the text itself will take. Thus, the text is stripped of claims of objectivity and the author's influence is exposed. For example, in carrying out scientific research and writing scientific research reports, standard, supposedly objective, scientific conventions are followed in terms of the way that both the research and the subsequent research report is structured. Deconstructive approaches trouble this concept, arguing that in writing and reading science:

> the science writer buries the subjectivity of the writer underneath the heavy prose of methodology, allowing technical language and the figural gestures epitomizing science to take control of the text. The writer's deep assumptions about the nature of the world are suppressed underneath the technical surface of the text, hidden from the community of science and thus protected from external challenges.
>
> (Agger, 1992, p. 102)

The third assumption identified by Agger in a deconstructive approach is that every cultural text is undecidable: the meaning is never given but is open to challenge and contestation. Thus, another assumption is that a deconstructive approach seeks the aporias, that is the blind spots, omissions, tensions, circumlocutions, and contradictions, in every text. Agger (1992) refers to these as the "internal fissures and fault lines" in a text (p. 102). A deconstructive reading pries open such fissures and fault lines to reveal the underlying subtextual frames of the text, some of which may actually be in competition with one another. Thus, another assumption of deconstructive approaches identified by Agger is that the subtext is turned into text. A deconstructive approach reads at the margins and views "many overlapping and cross-cutting texts—texts within, and beyond, texts, stories within, and beneath stories" (p. 108).

Finally, a deconstructive approach assumes that reading "writes," "because there is no way to develop readings outside of language and textuality" (Agger, 1992, p. 105). In so doing, deconstruction challenges the privileging of the written text over the reading of that text. Every act of reading any text creates a new and different text. Thus, the reading of any text and the criticism that accompanies it transforms that text, opening it to "new versions of itself by bringing to light its hidden assumptions and inconsistencies" (p. 96).

The Concept of Binary Oppositions

The notion of privileging the written text over the reading of that text highlights one of the very specific points in Derrida's deconstructive approach. This is the concept of binary oppositions, whereby one term is always prior or dominant to the other which is secondary or subordinate. For example, in the assumption of the authority of written text as opposed to the authority of the reading of that written text, there is a binary structure of meaning and value in operation. "Written" is the prior or dominant term, and "reading" the secondary or subordinate term. We can express this binary opposition as written/read.

Foucault (1982) also uses the idea of binary opposites and how we need to understand and critically analyse the relations of power that are associated in the construction of binary opposites. For example, Foucault offers oppositions such as "the power of men over women, of parents over children, of psychiatry over the mentally ill, of medicine over the population, of administration over the ways people live" (p. 780), as places where one might begin to look at how power operates. In all of these structures the first named term is the dominant term which is afforded primacy over the:

> secondary "weaker" or derivative term in the pair that is defined in terms of "not the dominant"... however, the definitional dynamic extends to the primary term as well in that it can only sustain its definition by reference to the secondary term. Thus, the definition and status of the primary term is in fact maintained by the negation and opposition of the secondary partner.
>
> (Cheek et al., 1996, p. 189)

The repression or deferral of the subordinate term is thus crucial for the governing position held by the superior term.

Derrida's thesis is that binary pairings are not "natural" or "normal": rather, they are constructions which reflect embedded assumptions about value and status. Thus, part of the deconstructive venture is to uncover such pairings and to expose their effects. In so doing, "binary oppositions thus become analytical sites of ongoing struggle and contestation" (Cheek & Rudge, 1994a, p. 19).

To illustrate the possibilities afforded by a deconstruction inspired approach, try reversing the order of the terms in some common binary pairings. What would the effect be of the following reversals: theory/practice to practice/theory; profit/non-profit to non-profit/profit; teacher/student to student/teacher; mind/body to body/mind; successful/unsuccessful to unsuccessful/successful?

Then ask yourself:

- Do these reversed binary pairings seem strange? Why/why not?
- What is being taken for granted?
- How is the primacy of certain terms and the subordination of others maintained?
- Who has an interest in such a maintenance of binary pairings?

These are important questions which can lead to many interesting and fruitful research undertakings. Such research challenges the very concepts and understandings that are at the core of many practices in applied contexts, yet they have often been taken for granted to the point that they are often assumed or are even invisible. Deconstructive approaches enable us to ask completely different sorts of questions in the research process. As Spivak (1976) asserts, in doing this type of research the text, which can be any representation of aspects of reality, "becomes open at both ends. The text has no stable identity, stable origin...each act of reading the 'text' is a preface to the next" (p. xii).

Have a look at a video developed by Megan as she talks about power and binary opposites https://www.facebook.com/megan.aston.12/videos/10218133504271154/

Deconstruction: Exploring Binary Oppositions in Practice

How might we use the ideas of binary opposition and deconstruction in practice? We answer this question using two examples.

Example One – Physical/Emotional Care

Aston et al. (2014) conducted a study that used a poststructural approach to deconstruct the 'text' or 'experiences' of children with intellectual disabilities, their parents, and nurses who cared for them in the hospital. Interviews with children, parents, and nurses provided the data or text that was then analysed. Deconstruction provided a lens to understand how binary opposites had been constructed through everyday beliefs and practices of nurses and families. Stories shared by participants demonstrated their frustration with a dominant medical discourse that prioritized physical care over emotional care. The binary opposites of physical and emotional care created a tension for all participants. Therefore, paying attention to the tension and marginalized practice of emotional care was the first step in deconstructing the experience of participants.

For example, many mothers spoke about how they did not have the opportunity to develop relationships with nurses and doctors and that if there had been more time for relationship building, their experiences would have been more positive. In addition

> Many nurses described situations that demonstrated how institutionally constructed patient care interfered with social/psychological interactions. In other words, high workloads had shifted care to more physical "tasks." Many of the nurses in the study spoke about workload interfering with building relationships. One nurse in particular stated that she believed spending time talking and playing with the children was just as important as medical tasks.
>
> (Aston et al., 2014, p. 228)

Example Two – Quantitative/Qualitative Research

Are there binary oppositions in place in terms of the values and emphases inherent in assumed understandings of research? Think of the debate that has surrounded qualitative versus quantitative approaches. Is a binary pairing in operation—namely quantitative/qualitative—in which quantitative approaches have been afforded primacy by many research "authorities" and qualitative approaches have been defined largely in terms of what they are not (Cheek, 2018, 2022)? Binary pairings such as quantitative/qualitative are not "natural" or "normal": rather, they are constructions which reflect embedded assumptions about value and status. Think also of the central binary pairing that has so influenced both research and claims to research "validity": objectivity/subjectivity. The effect of such a dichotomy has been to privilege certain so-called neutral and value-free approaches to research. However, as we have seen, claims to being neutral and value-free can hide assumptions about world views and may operate to conceal other ways of viewing or researching the same reality. Research methodologies themselves are texts which are framed by, and in turn frame, understandings about the nature of knowledge and about certain views of the way to research any given aspect of reality (Guba & Lincoln, 1994).

This is not to deny the importance of research methods, including traditional scientific/empirical approaches. Rather, it is to explore the concepts and assumptions embedded in any methodology employed and to consider the effect that these assumptions have on the research undertaken, including the methods used and the way the research is reported. Such a deconstructive project opens up many possibilities:

> Methodology can be read as rhetoric, encoding certain assumptions and values about the social world. Deconstruction refuses to view methodology simply as a set of technical procedures with which to manipulate data. Rather, methodology can be opened up to readers intrigued by its deep assumptions and its empirical findings but otherwise daunted by its densely technical and figural nature.
>
> (Agger, 1991, p. 114)

Thus, deconstructive approaches are effective for "interrogating taken-for-granted assumptions about the ways in which people write and read science" (Agger, 1991, p. 106). Such interrogation is crucial if other approaches are to be enabled to surface. As Agger points out, many published research reports:

> rely on the rituals of methodology in order to legitimate a certain form of knowledge. In these formulaic journal articles, methodology is not written or read as the perspectival text it is.
>
> (p. 122)

Consequently, the idea of research *itself* and the way that research is represented become "analytical sites of ongoing struggle and contestation" (Cheek & Rudge, 1994a, p. 19).

It is important to recognize that the intent of such deconstructive analyses is not to unpack, unravel and in the process destroy everything and leave nothing in place. Nor is deconstruction anti-science or anti-method as critics such as Kuntz (2012) would have us believe when claiming that such "thought is being used to attack the scientific worldview and undermine scientific truths; a disturbing trend that has gone unnoticed by a majority of scientists" (p. 885). It does not seek to reverse binaries in order to give priority to, or privilege, yet another set of terms e.g. qualitative/quantitative; or subjective/objective. Rather in a deconstructive approach all terms are contested and are constantly open to scrutiny and challenge. Hence, deconstructive approaches expose deep assumptions embedded in practice, and enable analyses that are able to be used to explore, contest and thereby possibly change, aspects of contemporary practice and practice settings.

Rounding Off

To round off this very brief introductory discussion of deconstruction, we wish to make two concluding comments. The first is that it is important not to give a sense of having provided "the" definition or understanding of deconstruction. All texts, including this one, are open to deconstruction: "every deconstruction can be deconstructed" (Agger, 1991, p. 115). All authorial privilege is contested in order to expose the investments any author has made in the text they produce, including in research - based texts. What an author has put in and what they have left out of any text, written or spoken, reflects such investments

The second point is that deconstruction is just as important and relevant today as it was when Derrida was writing about it decades ago. For example, both Vaittinen (2020) and Trumbull (2022) write about the importance of using Derridian deconstruction to understand humanness and life in general. Indeed, Trumbull is concerned that we have forgotten about the usefulness of deconstruction and it is necessary to show "that deconstruction does speak to resolutely material issues, and to life and the basic structure of the living." (p. 1).

Similarly, Vaittinen (2020) argues that there is a need for practice areas involved in care to examine how caring is both political and moral and what aspects of care are being silenced. This requires us

> to return to the philosophy of deconstruction, as devised by Jacques Derrida. In short, deconstruction denotes the practice of a systematic tracing of silences and suppressed meanings in texts. It involves a double move, where one first works with the dichotomies on which meaningfulness relies to undo and displace their hierarchical opposition so that in the second move of deconstruction, the terms of these hierarchical binaries can eventually be situated anew.
>
> (Par 21)

In the next chapter, we take a detailed look at discourse analysis as another way of putting postmodern and poststructural thinking into practice.

References

Agger, B. (1991). Critical theory, post structuralism, post modernism: Their sociological relevance. *Annual Review of Sociology*, *17*, 105–131.

Agger, B. (1992). *Cultural Studies as Critical Theory*. London, Falmer Press.

Aston, M., Breau, L., & MacLeod, E. (2014). Understanding the importance of relationships from the perspective of children with intellectual disabilities, their parents, and nurses *Journal of Intellectual Disability*, *18*(3), 221–237. https://doi.org/10.1177/1744629514538877

Ball, S. (1990). Introducing Monsieur Foucault. In S.J. Ball (Ed.), *Foucault and Education: Disciplines and Knowledge*. London, Routledge, pp. 1–8.

Cheek, J. (2018). The marketisation of Research: Implications for qualitative inquiry. In I. Denzin, K. Norman, & Y.S. Lincoln (Eds.), *The Sage Handbook of Qualitative Research* 5th edition. Sage Publications pp. 322–340.

Cheek, J. (2022). The impact of funding on ways qualitative research is thought about and designed. In *The SAGE Handbook of Qualitative Research Design*, pp. 636–651.

Cheek, J., & Rudge, T. (1994a). Nursing as textually mediated reality. *Nursing Inquiry*, *1*(1), 15–22.

Cheek, J., & Rudge, T. (1994b). Webs of documentation: The discourse of case notes. *Australian Journal of Communication*, *21*(2), 41–52.

Cheek, J., Shoebridge, J., Willis, E., & Zadoroznyj, M. (1996). *Society and Health: Social Theory for Health Workers*. Melbourne, Longman Cheshire.

Denzin, N. (1994). Postmodernism and deconstruction. In D. Dickens & A. Fontana (Eds.), *Postmodernism and Social Inquiry*. New York, Guilford, pp. 182–202.

Derrida, J. (1976). *Of Grammatology*, Gayatri Chakrovorty Spivak (Trans.) Baltimore, John Hopkins University Press.

Fairclough, N. (1992). Discourse and text: Linguistic and intertextual analysis within discourse analysis. *Discourse and Society*, *3*(2), 193–217.

Feldman, M. (1995). *Strategies for Interpreting Qualitative Data*. Thousand Oaks, Sage.

Foucault, M. (1982). Afterword: The Subject and Power. In H. Dreyfus & P. Rabinow (Eds.), *Michel Foucault: Beyond Structuralism and Hermeneutics*, Chicago, University of Chicago Press, pp. 208–226.

Guba, E.G., & Lincoln, Y.S. (1994). Competing paradigms in qualitative research. In N.K. Denzin & Y.S. Lincoln (Eds.), *Handbook of Qualitative Research*. SAGE, pp. 105–117.

Koro-Ljungberg M., Douglas E.P., Carlson D., & Therriault D.J. (2015). An unfinished dialogue about problematizing knowledge production in the peer review process. In N.K. Denzin & M.D. Giardina (Eds.), *Qualitative inquiry and the politics of research*. Walnut Creek, CA, Left Coast Press, pp. 27–50.

Kuntz, M. (2012). The postmodern assault on science. If all truths are equal, who cares what science has to say? *EMBO Reports*, *13*(10), 885–889. https://doi. org/10.1038/embor.2012.130

Mease, J.J. (2017). Postmodern/poststructural approaches. In *The International Encyclopedia of Organizational Communication*. Craig R. Scott & Laurie Lewis (Editors-in-Chief), James R. Barker, Joann Keyton, Timothy Kuhn, & Paaige K. Turner (Associate Editors). John Wiley & Sons. https://doi.org/10.1002/9781118955567.wbieoc167

Nicholls, D.A., & Cheek, J. (2006). Physiotherapy and the shadow of prostitution: The Society of Trained Masseuses and the massage scandals of 1894. *Social Science & Medicine*, *62*(9), 2336–2348.

Norris, C. (1991). *Deconstruction, Theory and Practice* (Rev. ed.). London, Routledge.

Spivak, G.C. (1976). Translators preface. In Gayatri Chakrovorty Spivak (Trans.) *Of Grammatology*, Baltimore, John Hopkins University Press.

Squier, S. (1993). Representing the reproductive body. *Meridian, 12*(1), 29–45.

Trumbull, R. (2022). *From Life to Survival. Derrida, Freud, and the Future of Deconstruction.* Fordham University Press.

Vaittinen, T. (2020). Exposed to violence while caring: From caring self-protection to global health as conflict transformation. In *Teoksessa Gender Global Health and Violence: Feminist Perspectives on Peace and Disease.* London & New York, Rowman and Littlefield International, pp. 227–250.

Weedon, C. (1987). *Feminist Practice and Post Structuralist Theory.* London, Basil Blackwell.

5 Discourse Analysis

One Way of Using Poststructural Approaches in Practice

In this chapter we will

- Explore discourse analysis, as a research approach often used in research drawing on poststructural approaches.[1]
- Examine how discourse analysis has been used in studies of practices in applied areas.
- Use examples to highlight the way that textual representations of practice are shaped by, and in turn shape, dominant discourses
- Provide some guiding principles to assist you in implementing discourse analysis

---

What Is Discourse Analysis?

Over two decades ago Julianne noted that although discourse analysis had gained more prominence in qualitative research studies in a range of applied areas, this had not "necessarily led to better understandings and/or use of discourse analysis as a research approach in qualitative research. If anything, the waters have become muddier rather than clearer" (Cheek, 2004, p. 1140). She noted such muddiness often arises from a lack of specificity as to what understanding of discourse and discourse analysis was being used in reports of studies claiming to use this research approach.

There are different theoretical and disciplinary views about the meaning of both discourse and discourse analysis. The definition of discourse, and therefore discourse analysis in play at any one time, reflects its theoretical and disciplinary underpinnings. For example,

> (D)iscourse analysis in sociolinguistics, sociology and social psychology, to mention just a few possibilities, are likely to differ in the sources

DOI: 10.4324/9780429053764-5

they refer to, and also, to some extent, in the problems and research questions which they set out to investigate.

(Taylor, 2013, p. 1)

Consequently, discourse analysis "does not refer to a single approach or method ... (it) refers to a range of approaches in several disciplines and theoretical traditions" (Taylor, 2013, p. 1). Discourse analysis is an inter-disciplinary concept drawing on linguistics, cognitive psychology, anthro-pology, sociology, and cultural studies, and is used in a variety of ways (Taylor, 2013). Different groups of researchers may understand discourse analysis in different ways. Hence, because of its interdisciplinary origins, discourse analysis will always be an approach with different emphases and understandings in use depending on the understandings of the researcher employing the approach. This is why

- "there is not a single or unitary set of rules for "doing" discourse anal-ysis" (Cheek, 2004).
- "it is possible to design discourse analysis in very different ways" (Cheek & Øby, 2023, p. 108).
- "there cannot be "the" set of rules for discourse analysis" (Cheek, 2004, p. 1148).
- the analysis of discourse(s,) or discourse analysis, "is not, or should not be, a "method" to be wheeled on and applied to any and every topic" (Parker, 1992, p. 122).

However, "although it is not possible to identify "the" set of rules for discourse analysis" (Cheek, 2004, p. 1148), this in no way suggests that discourse analysis is a "free-for-all" where anything goes and anything can masquerade as discourse analysis. As Van Dijk (1997) indicates, all approaches to discourse analysis involve rigorous methods and principles of "systematic and explicit analysis" (p. 1).

An analysis of discourse is a scholarly analysis only when it is based on more or less explicit concepts, methods or theories. Merely making "commonsense" comments on a piece of text or talk will seldom suf-fice in such a case. Indeed, the whole point should be to provide insights into structure, strategies or other properties of discourse that could not readily be given by naive recipients.

(p. 1)

Stephanie Taylor (2013) in her excellent introductory book to what dis-course analysis is, devotes her first chapter to demonstrating "the vari-ety of discourse analytic approaches and also of the kind of research problems they have been used to address" (p. 5). Using 4 published journal articles she demonstrates how they vary both in the approach to discourse analysis that they use and the "kinds of research problems they have been used to address" (p. 5). She gives a useful "overview of each research project, its theoretical grounding, the empirical work and the data which were analysed, and the discourse analytic or discourse approach which the researcher has adopted" (p. 5).

Similarly, in his book Doing Discourse Research: An introduction for social scientists, Reiner Keller (2012) introduces the basic principles of discourse analysis and in Chapter 2 pp. 13–32 explores different ways discourse is thought about and put into practice.

Poststructural Inspired Approaches to Discourse Analysis

Discourse analysis arising from, and underpinned by, poststructural thought is informed by the "notion of language as a meaning constituting system which is both historically and socially situated" (Cheek & Rudge, 1994b, p. 59). Texts,[2] whether they are books, articles, newspaper reports, interviews, observations, or drawings, are embedded within discursive frameworks. They are constructed by the understandings of particular discourses[3] and in turn they construct understandings in keeping with those discursive frames. Consequently, "(M)eanings, as they occur in... text[s] are the product of dominant discourses that permeate those texts. Not only do powerful discursive frameworks provide meaning for the text, they actually frame the text itself in the first place" (p. 61).

Approaches to discourse analysis inspired by poststructural thinking "therefore generally involve:

- some form of textual analysis
- some sort of structured investigation of the broader discourse of which the focal texts are a part
- and an investigation of the social context in which the texts appear melded together to produce insights into the social world."
  (taken from Phillips and Di Domenico, 2009, p. 551 – bullet points added by us)

These discursive analyses of texts are thus not simply descriptions or analyses of content. Rather, they are critical and reflexive, moving beyond the level of common-sense. The analysis focuses on "meaning constituted systems" (Cheek & Øby, 2023, p. 107) in texts generated from, for example, interviews, news articles, or visual texts such as pictures and films (Taylor, 2013). "(D)iscourse researchers are always looking for succinct data extracts which *exemplify* patterns that recurred across an entire data set" (Taylor, 2013, p. 75).

Such critical and reflexive analyses situate texts in their social, cultural, political, and historical contexts. Questions that may be asked of a text include

- "Why was this said, and not that?," "Why these words?," "Where do the connotations of the words fit with different ways of talking about the world?" (Parker, 1992, p. 4)
- "What could this be evidence of? What do I want to argue on the basis of this evidence?" (Taylor, 2013, p. 74)
- "How did these texts come to be the way that they are in terms of the meanings that they convey.... and what sustains the understanding that they produce?" (Cheek & Øby, 2023, p. 107)

Texts are thus interrogated to uncover the unspoken and unstated assumptions implicit within them which have shaped the text in the first place, *and* the effect of those assumptions.

Hence discourse analysis enables an exploration of textual representations in terms of the way they are shaped by, and in turn shape, dominant discourses – ways of thinking and talking about reality that demonstrate the inextricable link between power and knowledge (Foucault, 1972). Thus, the analysis moves beyond the descriptive into the discursive and enables the outcomes of the research to be much more than just a study of words – one of the critiques often levelled at discourse analysis (Taylor, 2013).

Put another way this view of discourse analysis

> views discursive activity as constitutive of the social world, not a route to understanding it... it is the structured and systematic study of collections of interrelated texts, the processes of their production, dissemination and consumption, and their effects on the context in which they occur.
>
> (Phillips & Di Domenico, 2009, p. 551)

In such an approach to discourse analysis the "text is not a dependent variable, or an illustration of another point, but an example of the data itself" (Lupton, 1992, p. 148). Texts *themselves* and how *they* came to be the way that they are, rather than just the reality they purport to represent, are the data to be analysed.

A good example of interrelated texts and competing discourses impacting on the way the reality that they purport to represent is the study by Arousell et al. (2017) which explored the tensions and uncertainties arising for health care workers in Swedish multicultural contraceptive counselling practice settings. These tensions arose from these health care workers attempting to navigate competing discourses of "gender equality promotion on one hand and respect for cultural diversity and individualized care on the other" (p. 1518). Tensions such as the "ultimate respect for individuals' cultural and religious beliefs would by definition mean that all values – also those that cherish *in*equality rather than gender equality- would be welcome" (pp. 1519–20). The focus of this study was to explore how health care workers in these practice settings navigated these tensions/competing positions and how this constructed practice in these multicultural contraceptive counselling practice settings.

You can read more about the study and the approach to discourse analysis that was used at

Arousell J, Carlbom A, Johnsdotter S, Larsson EC, Essén B. Unintended Consequences of Gender Equality Promotion in Swedish Multicultural Contraceptive Counseling: A Discourse Analysis. *Qualitative Health Research*. 2017;27(10):1518–1528.

Doing Discourse Analysis: How Might a Poststructural Inspired Approach to Discourse Analysis be Put into Action?

Although there is not a set method to poststructural inspired discourse analysis, there are some guiding principles that we can draw on when designing and conducting this type of research. For example, Parker (1992) outlines four key features, or stages, of such discourse analysis. However, the word stages should not give the impression that this is a sequential linear process. Rather, putting this type of discourse analysis into practice is an iterative process "involving cycles of thinking where you begin with an idea, think it through, and then revisit the initial idea that you had, refine or change it in line with that thinking, and then think that change through and so on." (Cheek & Øby, 2023, p. 2).

This includes thinking related to what the study is about (the questions/purpose), the texts chosen for analysis, the ways to collect or obtain those texts, and how those texts are analysed. For, as we have seen, there are many types of texts. Such iterative thinking about research into discourse "should be led by the issues and problems that are to be addressed and, where possible, by research participants" (Parker, 1992, p. 122).

With these points in mind, the four stages/features of discourse analysis posited by Parker are as follows:

1 Introduction.
  This is where the study is positioned with respect to its relationship to other works in the substantive area. These works are drawn from a "traditional" search of the literature, as they are in any other research undertaking, and do not include only studies using discourse analysis. Further, the type of texts to be analyzed as well as the types of questions/issues driving the research are discussed in order to contextualize the research.

2 Methodology.
  In this section detail is given about the specific texts to be analyzed. Why these texts and not others were chosen as the focus of the study is an important consideration to be discussed here. Information is also given about how these texts will actually be obtained. For example will it be by interviews, observations, collecting forms of written or visual texts, and/or all/some of the above. As in any research plan, detail is needed about the type of interviews and/or observations conducted, and/ or about the type of written and visual texts collected.

3 Analyses.
  Parker outlines the coding of the data (which, in discourse analysis, is the text itself) under different discourse headings. Of particular interest to the analysis is any absence of possible discursive frames. That is, what ways of speaking and thinking about the reality in question are not present and why might that be so? There is no set way of doing such analysis and Parker, suggests that it is inevitable that a degree of intuition must be employed.

4 Discussion.
  In this section the analyses are linked to other material in the area in order to draw out points of discussion about the substantive area under scrutiny. This section also involves "reflection on the issues raised by the method including, crucially in the case of material in which you participated (such as interviews), the position of the researcher."

(Parker, 1992, p. 123)

Parker's (1992) book also has an interesting section in Chapter 6 that looks at a range of areas and ways in which discourse analysis has been employed, and you may find it useful to read this. However, a key suggestion made by Parker is that "perhaps the best way to get a feel for forms of discourse is to look at how analysts actually deal with texts" (p. 127). This is sound advice. In the next section we will explore one published study that puts Parker's guiding principles for implementing discourse analysis into practice.

Putting Discourse Analysis into practice: Analysing Case Notes Discursively

Discourse analysis can be talked about and described *ad infinitum*, but examining actual studies greatly assists in demonstrating how discourse analysis may be used to, and its potential for, problematizing (Foucault, 1986) and thereby informing understandings of practices. With this in mind, we will explore an example of how discourse analysis was used to explore an aspect of practice.

The study that forms the basis for this example was carried out by Julianne and a colleague, Trudy Rudge (Cheek & Rudge, 1994a, 1994b).[4] Although the research was conducted several decades ago the example still provides clear insights into how a study using discourse analysis can be designed and put into practice drawing on Parker's guiding principles related to four stages/features of discourse analysis – Introduction, Methodology, Analyses, Discussion. Julianne outlines the process they undertook.

Introduction

We analysed the case notes that were developed about two patients in a rehabilitation setting. The case notes were examples of health texts that were organized by, and in turn helped to organize, the dominant discourses framing the practice setting in which they were developed. The analysis took as its focus the interaction between the text (the case notes) and the context in which the text existed. A basic premise of this study was that nursing and other aspects of health care are textually mediated and that discourse analysis can provide insights into how text "both constitutes and reproduces the social act of nursing" (Cheek & Rudge, 1994b, p. 61). In such analysis "our concern is not whether these are either a poor or good record of the events, but rather the nature of the reality produced by these texts" (Cheek & Rudge, 1994a, p. 43). Such a focus moved the research beyond the descriptive and located it in the realm of the discursive. Put another way, the focus is not so much on what is recorded as it is on why it is recorded, and conversely, why other things are not recorded. How did, and do, case notes come to be developed in the form they are, and why and how is such a form maintained?

Our broad research goals were to look at how the case notes and the accompanying "case" (that is, the patient) were represented textually in the notes. Case notes are "a deliberately 'crafted' document" (Poirier & Brauner, 1990, p. 30) and reflect the influence of often competing discursive frames such as managerial, legal, and medical discourses. Further, we wanted to look at how the actual space in the case notes was used or not used, and how it was managed by the various groups of health care providers writing the notes. The questions of who wrote, when they wrote, what they wrote about, and how they wrote were of great interest to us as this analysis offered important insights into how the groups of care providers interacted with each other and the patients in the webs of documentation developed. As Poirier and Brauner (1990) point out, although medical records (or in this study case notes) are constructed by diverse voices, "their notes are entered into the chart in a prescribed way, producing an organized record with a familiar sequence of information and events" (p. 30).

Methodology

How was this research actually done? The case notes of two randomly selected patients in a rehabilitation setting were analysed. A rehabilitation setting was chosen because length of stay in this type of setting tends to be longer than in other health care settings; thus, it could be expected that the documentation and case notes developed about each patient would be substantial. In fact, both patients' notes in the analysis covered stays of four months. Further, by its nature, the rehabilitation setting is multidisciplinary. Consequently, the notes developed reflect input from a large number of health professionals, thereby providing insights into the ways each professional writes about their practice independently but also how that practice and what they write about it relates to the other health care providers in the rehabilitation context. All relevant ethical clearances were gained before the research commenced, including the consent of the two patients. The patients were Mrs. W, a 79-year-old woman who had fractured her femur in a fall, and Mr. H, a 63-year-old man who had suffered a stroke while on holiday.

Analyses

The constraints of space in a book such as this do not allow for an in-depth discussion of each patient's case notes and their analysis. Given such constraints, what follows is an illustration of some of the key points to emerge from this research from which we could give a sense of what type of analysis might emerge from such a study. The study has been reported in its entirety elsewhere (Cheek & Rudge, 1994a, 1994b).

The analysis began by looking at five specific areas. These were:

- the format of each entry: typed, handwritten or computer-generated;
- the ways in which the individual writers identified themselves: were entries signed or not, and whether titles or qualifications were given;
- the similarities and differences between the various groups of health professionals in terms of content – what they wrote about and, of even more interest at times, what they didn't write about;
- how each group of professionals used the space in the notes – whether they wrote profusely, succinctly, or not at all;
- the form of language used: for example, was it the language of objective facts, using notations and symbols drawn from the discursive frame of the scientific/medical?

Each set of case notes revealed a common sequence of events in the initial construction of the notes. There was a record of what we termed the "rite of admission" for each patient in their notes. Such a rite was clearly situated in the frame of medical/scientific discourse where the "truth" of the patient's condition was established by examination by experts. In so doing, these experts used the authority given to them by scientific/medical discourse to establish the truth about the patient's condition. As a result of this examination, the patient can be located within such discourse— hence Mrs. W was described as "a fall who fractured her femur."

Once the patients were located, they were then able to be considered object of, and subject to, the medical/scientific discourse that defined them in the first place. Admission was thus concluded by the issuing of an identity number to each patient by which they could be monitored and scrutinized, and by the attaching of a wrist band to convey and confirm the identity of each patient. Further, additional statistics were collected about each patient to contribute to public health statistics and to meet government requirements for information about admissions and funding. At this point, the patients made their first, if not only, appearance in the notes. A record of consent given by each patient for treatment was found in each of the case notes following documentation regarding admission. However, there was no information about what each patient had been told about their condition or about what they had based their consent on.

Analysis of the notes reflected the centrality of the examination in contemporary health care provision. Each professional conducted their own examination of each patient and each group of experts used the examination(s) to situate the patient within their own field of expertise. At times, it was as though each professional was almost oblivious to what other care providers were doing, and the voices of each professional group seemed to speak "past one another, barely acknowledging each other's existence" (Cheek & Rudge, 1994a, p. 45). In this respect, the case notes were comparable to Bakhtin's (1981) description of a novel:

"a diversity of social speech types...and a diversity of individual voices artistically organized" (p. 262). Poirier and Brauner (1990) also used literary metaphors to describe the construction of the case report in terms of the various contributions to the report:

> The case report might more appropriately, in fact, be read as an anthology of short stories, a collection of world views which are brought together to address the same topic but which reflect the unique views of their individual authors.
>
> (p. 38)

Thus, it is not surprising that at times it is not clear who the audience is, for either the case notes in general, or specific parts of the notes.

Discussion

It was apparent that the notes assumed a privileged position in the health care setting. They were available only to "insiders." Not even the patient was openly granted the right to see these notes: the ownership of the notes was clearly with the health professionals. Further, the mystique associated with the symbols, abbreviations, jargon, and shorthand used by exponents of medical/scientific discourse within the notes confirmed the exclusion from the notes of all but those privy to such knowledge. The patient as person effectively disappeared, irrelevant to such discursively framed notes. The patient was present only as defined "in relation to the contours of medical knowledge and practice" (May, 1992, p. 591). The patient was thus reported as object: the source of information to be gleaned from examination. Further, the patient was subject to conclusions drawn from examination.

Nowhere was this more evident than in the following entries from Mrs. W's notes:

> *Claims* [italics added] can't see out of right eye.... *claims* [italics added] she also struck occipital region of head.... *claims* [italics added] this is because of deteriorating R eye.... Viewpoint-no definite cause for falls.
>
> (Cheek & Rudge, 1994a, p. 47)

Mrs. W's intimate knowledge of her own sight and her experience of her falls were discounted. The detached objective conclusion, following some sort of examination, was that there was no definite cause for her falls. As we have pointed out, in this instance the patient was:

> constituted by the perceptions of the various "experts" recounting their perceptions of reality. Furthermore, these case notes [were] taken as objective, factual records of the patient's progress, whereas in actuality they represent[ed] one version of a number of possible realities.
>
> (Cheek & Rudge, 1994a, p. 48)

The patient, being effectively excluded from access to what was written in the notes, had no right of reply.

The subjects/patients must then comply with regimes of treatment for their own good in the quest to restore "normality" according to the norms established by medical/scientific discourse. Such compliance is part of the development of the docile patient: "good" or "bad" patient can be gauged by the extent of conformity with directives.

In terms of "conspicuous absences" in the notes, an analysis of the nursing documentation proved enlightening. Nurses tended to report the day's events not as they themselves had experienced it but in terms of the discourse of others, predominantly the medical/scientific.

Often it was others who reported important information drawing on nurses' experiences. For example, the following entry was made by a speech therapist in Mr. H's notes: "Ward staff have described his wife's considerable emotional difficulties, and it is clear that a certain lack of rapport currently exists between them" (Cheek & Rudge, 1994a, p. 49).

Therefore, questions that arose related to these conspicuous absences included: Why did the nurses, who had made the important observation above about Mr H and his wife in the first place, not record it? Why was it left to the speech therapist to record this, and not to the group of health care providers who have the most constant contact with the patient and his family?

The analysis of these two sets of case notes, with respect to the nursing voice and to voices within that nursing voice supports Parker and Wiltshire's (1995) contention that:

> because of the dominance of the medical voice in the hospital context, supported as it is by the disciplinary power of medical knowledge, it is not surprising that the nursing voice is somewhat muted.
>
> (p. 166)

However, possibly even more surprising is how discourse analysis of case notes reveals that nurses are actually complicit in such a muting or silencing of aspects of the nursing voice!

> It is not just that the… [case notes]…*represent* nursing as a process that is technical in its intent, but that these practices are *constituted* by the taken-for-granted nature of the authority embedded in both the content *and* form of such texts.
>
> (Cheek & Rudge, 1994b, p. 66)

To sum up

This study is an exploration of how case notes have been shaped predominantly by certain discourses and especially by scientific/medical discourse. From this study it is possible to see the potential that discourse analysis has to offer analyses of specific procedures in health care and health care practices. The study illustrates the type of data and analysis that might form part of a discourse analysis of the documentation that is routinely developed about patients in health settings. In so doing, the study demonstrates how discourse analysis can be used to explore and expose health care practices as textually mediated. This opens up other ways of viewing those practices.

Tips for Getting Started with Discourse Analysis

When introducing poststructural inspired discourse analysis to her graduate research classes Megan developed some guiding principles to help students struggling with knowing where to begin such discursive analysis. These are:

1 Identify important issues.
   Read/view the 'text' (transcript, video, pictures) and mark moments, words, or quotations/images you feel represent important issues that are related to the research topic or question you are examining.

2 Apply concepts beliefs, values, and practices.
   Referring to each marked section of the text identify what beliefs, values and practices are reflected by the words or images of each of the marked sections.

3 Develop social and institutional discourses.
   Write about the social and institutional discourses you see informing the issue you identified. Sometimes this is clearly described in the quotation or evident in a visual image, but most often you need to expand on the implied ideas. Try to identify any discursive framework that seems to be shaping the text or image you have highlighted. What social and/or institutional meanings contribute to this discourse?

4 Identify relations of power.
   Referring to the parts of the interview texts or pictures or videos you have highlighted write about the discourses you see the participants connecting with or that are impacting participants' experiences. How do the discourses affect the participant? Do the participants agree or disagree with social and/or institutional beliefs, values, and practices? Is it an easy or positive fit? Or are there questions, conflicts, tensions? These are the "relations of power" that the participants are feeling/experiencing.
   Pictures or videos: write about the discourses you see connected to visual imagery.

5 Identify subjectivity and agency.
   Consider how the text represents the participant's "subjectivity" (how they are positioned as a nurse, man, woman, teacher etc.) as well as their "agency" (how they choose to act in each situation by fitting in or challenging other discourses). Agency is how they use their power.

Visual imagery and text may also present different aspects of agency and power.

You can read more about these 5 guiding principles in Megan's article

Aston, M. (2016). Teaching feminist poststructuralism: Founding scholars are still relevant today. *Creative Education*, 7(15), 2251–2267. doi:10.4236/ce.2016.715220

Putting Megan's Guiding Principles into Action Using the Music Video 'Take on Me'

Megan then uses a text in the form of a music video clip to demonstrate how to put these guiding principles into practice. We will use the first musical clip Megan used with a class as the vehicle for the discussion here. It is the 1986 music video Take on Me by the Norwegian group 'Ah ha.' https://www.youtube.com/watch?v=djV11Xbc914 Before reading further please view the video.

The task for the class is to identify how the beliefs, values, and practices are reflected by the words or images in this music clip. After they have viewed the video clip the class works together to identify the text(s) of the song 'Take On Me' which includes visual images of real people and cartoon characters, as well as the music and words. Students can also find the written words on the internet.

Students quickly identify the historical context of the music clip and try to understand the beliefs and values that were perpetuated in the 1980s through personal experience if they are old enough, or through media or written text. When doing so students are reminded that although they may have their own preconceived ideas about these beliefs and values, it is important that they focus closely on the exact words and presentation of meanings and beliefs that are in the video. This helps them focus on the text and begin to recognize their own biases and judgements. For example, some social beliefs can be seen in the social constructions of the characters including, 'tough boys,' 'sweet girls,' boys running from the law, boys expecting girls to take care of them, girls running after boys, young love.

After identifying some social beliefs and values, we look carefully at the words, imagery, and actions in the video to see how these social ideas are perpetuated and create meaning through different discourses. Do the social beliefs and values presented in the video resonate with our understanding of the world in the 1980s and are there connections with present day? This enables us to begin discussing possible discourses that might have influenced the writing of the song and presentation of the music video and images. For example, we identify, discourses on 1) heteronormative gender expectations 2) stereotypical body images and 3) predominant representation of white race.

Although the discourses are often explored individually initially, the relations/intersections between different discourses and how people react to discourses and are affected by them show how power is negotiated. For example, binary heteronormative gender relations and tensions are seen in the images and words between the young female and young male. The male is asking her to 'take him on' and 'take a chance' but he seems to be running from the law. He also says he will be 'coming for her' a relation that could/could not analysed as inappropriate behaviour. We also identify subjectivity in the form of imagery including idealised body shape for main characters such as small body size and stereotypical beautiful young people, as well as stereotypical/stigmatized bodies for the large waitress and angry looking police – and all characters are white.

We discuss how practices and 'choices' that are made by the characters can be understood as one's agency in the way that they 'negotiate' their roles and how they act in the video. For example, the young woman follows the young man into his comic-based world along with the police. With all of these intersections, we can see how the socially constructed discourses of gender, body, and race contribute to the way the characters use their agency to negotiate relations of power in the music video.

It is also important to consider the music and how it contributes to the construction of discourses. An uplifting, fast, and frantic beat can be interpreted as creating a discourse of fun and heterosexual connections; however, there is also a dichotomy when positioned against tensions of running from the law and 'bad boy' images. We can see the complex construction of different discourses and how all of these identified social beliefs create meaning about heterosexual relationships in the 1980s.

This activity is meant to be a fun, compact, and engaging way to demonstrate how discourse analysis inspired by poststructural thinking can be used to explore aspects of our everyday lives and practices.

When Megan first introduced this exercise, one of the students found a video that 'challenged' the 'Ah ha' music video by using different imagery including a lead Black young female, older balding man, and a slim waitress. One can quickly see the shift and challenge to dominant social discourses related to the way gender, race, and the body are constructed and what is viewed as "normal" and desirable, and what is not. Check it out: https://www.youtube.com/watch?v=J6jHqvc--Nw

Rounding Off: Some Inspiration Tempered with Words of Caution

Far from being removed from understandings of practice and therefore unable to contribute to them, discourse analysis challenges the everyday, often taken-for-granted realities of practice. However, like any research technique and approach, discourse analysis has limitations. As we have seen, discourse analysis draws on traditions from many disciplines, each of which has its own particular emphasis on the way discourse analysis is operationalized. Because it allows for multiple perspectives, such diversity can be a strength of discourse analysis: but it can also be a limitation if the approach used is poorly defined or not contextualized in terms of its theoretical origins. In addition, it is important that discourse analysis does not remain only at the micro-level of analysis but is also extrapolated to the macro level. It must consider the social and political realities of the context by which the text is mediated and which in turn it mediates.

Finally, it is important to point out that readers of texts are not passive consumers of those texts: rather, individuals are active readers of texts. As Gledhill (1988) points out, "Reading is itself an active, though not free, process of construction of meaning and pleasure, a 'negotiation' between texts and readers whose outcome cannot be dictated by the texts" (p. 113). Every textual reading (including this one) is a negotiated one in the sense that there is a negotiation on the part of the reader between the viewing position created by the text and the understandings that the reader brings to that text (Cheek, 1996). Yet, as Frazer (1992) points out, "all too often theorists commit the fallacy of reading 'the' meaning of a text and inferring the ideological effect the text 'must' have on the readers (other than the theorists themselves, of course!)" (p. 186).

Thus, research into the readers of texts is needed to investigate the negotiated readings of those texts that are actually adopted. Discourse analysis of texts is somewhat speculative if the way texts are read is not investigated. Nevertheless, discourse analysis remains a useful and important research approach that is able to expose and frame the reality of our contemporary world. It is thus a useful way to explore the paradox that exists with respect to these texts:

> The paradox is that not only can textual portrayals and analyses of health care practice enhance our understandings of health, they can also limit them if we do not recognise that the viewing position adopted, from which to frame our analysis, can have the effect of confining our understandings to certain parameters.
>
> (Cheek, 1995, p. 60)

Coda: Is This All Old Hat?

Finally, a word to those who may be thinking that we have moved on from discourse analysis and/or discourse analysis may be past its expiry date based on the fact that the use of the term discourse analysis is not as common in social sciences or the research methodology literature as it once was. We do not agree. Often the principles and ideas of discourse analysis have been used to inform research that is called something else. As Taylor (2013) points out,

> (W)ithin social psychology, for example, research in the areas of discursive psychology and rhetorical psychology is a development from the sources which originally referred to discourse analysis. Other social researchers incorporate an analysis of discourse into a differently named research approach, such as ethnography or psychosocial research.
>
> (p. 85)

Such observations take us back to where we began this chapter namely that discourse analysis is not a set or fixed method – a "thing" – as such. Rather it is a way of thinking that can be applied to research about textual representations of reality in many different ways – including as part of a study using a combination of methodological or theoretical approaches.

The remaining chapters of the book focus on practical or 'how to' aspects of employing postmodern and poststructural approaches in research. The next chapter provides tips about how to write a proposal for research that uses postmodern and poststructural approaches.

Notes

1 In keeping with the remit and focus of the book the discussion is limited to examples where the discourse analysis employed draws on post structural and postmodern thought.
2 See the discussion of text and textually mediated reality in Chapter 4.
3 See the discussion of discourse in Chapter 3.
4 We introduced this study in Chapter 4 when introducing the idea of texts and textually mediated reality.

References

Arousell, J., Carlbom, A., Johnsdotter, S., Larsson, E.C., & Essén, B. (2017). Unintended consequences of gender equality promotion in Swedish Multicultural Contraceptive Counseling: A discourse analysis. *Qualitative Health Research*, *27*(10), 1518–1528. https://doi.org/10.1177/1049732317697099

Aston, M. (2016). Teaching feminist poststructuralism: Founding scholars are still relevant today. *Creative Education*, *7*(15), 2251–2267. https://doi.org/10.4236/ce.2016.715220

Bakhtin, M. (1981). *The Dialogic Imagination: Four Essays by Mikhail Bakhtin*. Austin, University of Texas Press.

Cheek, J. (1995). (Re)viewing health: Textual representations of aspects of contemporary Australian Health Care. *Australian Journal of Communication*, *22*(3), 52–62.

Cheek, J. (1996). Taking a view: Qualitative research as representation. *Qualitative Health Research*, *6*(4), 459–507.

Cheek, J. (2004). At the margins? Discourse analysis and qualitative research. *Qualitative Health Research*, *14*(8), 1140–1150).

Cheek, J., & Øby, E. (2023). *Research Design. Why Thinking about Design Matters*. SAGE.

Cheek, J., & Rudge, T. (1994a). Webs of documentation: The discourse of case notes. *Australian Journal of Communication*, *21*(2), 41–52.

Cheek, J., & Rudge, T. (1994b). Inquiry into nursing as textually mediated discourse. In P. Chinn (Ed.), *Advances in Methods of Inquiry for Nursing*, Gaithersburg, Aspen Publishers, pp. 59–67.

Foucault, M. (1972). *The Archeology of Knowledge and the Discourse on Language*. New York, Pantheon Books.

Foucault, M. (1986). Kant on enlightenment and revolution. *Economy and Society*, *15*(1), 88–96.

Frazer, E. (1992). Teenage girls reading "Jackie." *Media, Culture, Society*, *9*, 407–425.

Gledhill, C. (1988). Pleasurable negotiations. In E. Pribham (Ed.), *Female Spectators: Looking at Film and Television*. London, Verso, pp. 64–89.

Keller, R. (2012). *Doing Discourse Research: An Introduction for Social Scientists*. Sage.

Lupton, D. (1992). Discourse analysis: A new methodology for understanding the ideologies of health and illness. *Australian Journal of Public Health*, *16*(2), 145–150.

May, C. (1992). Individual care? Power and subjectivity in therapeutic relationships. *Sociology*, *26*, 589–602.

Parker, I. (1992). *Discourse Dynamics: Critical Analysis for Social and Individual Psychology*. London, Routledge.

Parker, J., & Wiltshire, J. (1995). The handover: Three modes of nursing practice knowledge. In G. Gray & R. Pratt (Eds.), *Scholarship in the Discipline of Nursing*, Melbourne, Churchill Livingstone, pp. 151–168.

Phillips, N., & Di Domenico, M.L. (2009). Discourse analysis in organizational research: Methods and debates. In D.A. Buchanan & A. Bryman (Eds.), *The Sage Handbook of Organizational Research Methods*, London, SAGE Publications Ltd, pp. 549–565.

Poirier, S., & Brauner, D. (1990). The voices of the medical record. *Theoretical Medicine*, *11*, 29–39.

Taylor, S. (2013). *What Is Discourse Analysis?* London, Bloomsbury.

Van Dijk, T. (1997). Editorial: Analysing discourse analysis. *Discourse and Society*, *8*(1), 5–6.

6 Proposing Research Informed by Postmodern and Poststructural Approaches

In this chapter we will

- Demonstrate that writing a research proposal is a craft
- Analyse and dissect a research proposal that used postmodern/poststructural thinking
- Highlight points to consider when proposing research using postmodern/poststructural approaches

Introduction: Research Proposal Writing as a Craft

Crafting a well-constructed research proposal requires skill. It is a craft and, like any craft, it improves upon practice. Developing a research proposal using postmodern and poststructural approaches requires both theoretical and methodological dexterity coupled with a touch of ingenuity to apply such theoretical and methodological frames to new or different substantive areas.

Often due to conventions and understandings about what is, or is not appropriate to report in scholarly articles, the finer details about how the research in question was designed is left out. Thus, after reading such reports one often finds it hard to work out exactly what was done, how, and why. Rarely, if ever, are research proposals from established researchers published so that the conceptual development of the research design can be explored.

Such invisibility of research proposals means less experienced and/or novice researchers seeking guidance in the art and craft of proposal writing, often find it hard to conceptualize what a research proposal using, for example, a postmodern approach might look like. The beginning researcher often has to rely on obtaining proposals from more experienced researchers and/or having discussions with them, which may not always be possible.

Therefore, in this chapter we aim to address this somewhat invisible nature of research proposal development. To do so, the discussion is

DOI: 10.4324/9780429053764-6

centred on analysing an actual proposal and how it was written. This is in contrast to most discussions in which general principles for proposal writing are identified and the assumption is made that the reader can go on to formulate proposals from such principles. Tips on how the reader can write a successful research grant proposal are woven into the discussion.

Getting Going: Formulating Research Aims and Questions

The first step in formulating any research proposal is deciding exactly what it is that will be studied. What is the issue/question/problem that will be the focus of the study? Ideas for research can arise from issues you confront in your everyday life, in your professional practice or practice setting, in what you have read in both professional and popular literature such as magazines and newspapers, in conversations with others, or in formal courses of study. See Chapter 3 in Cheek and Øby (2023) for a fuller discussion of this.

The development of the idea into a research proposal will require you to think about the theoretical and methodological frames to be employed in the study. What type of theoretical approach will you use to frame the research and why? What type of data will you collect and why? The type of data collected will depend on the theoretical and methodological frames employed.

Developing a question, topic, or issue into a researchable form takes time and energy. A thorough literature review will help refine the research topic as it can be used to reveal what other studies have been done in this area, what research techniques have been used, and whether or not your proposed research seems feasible. Once your question has been refined, it is possible to establish aims for the study and your methods for researching the particular area. However, until you have clearly refined the question or issue that drives your research, you will not be able to move any further in your proposal development. As simple as it may sound, identifying and formulating a question for the research is a complex skill integral to the craft of proposal writing. This includes research that is proposed using postmodern/poststructural perspectives.

Tips for developing your own research postmodern/poststructural topic

1   Choose an area of interest. What are you passionate about? What do you believe needs to be researched? What do you want further answers about? Is there something you would like to further question? Do you have a concern about a certain work or social practice? Would you like to highlight a practice that might be invisible/less recognized? Is there a 'taken-for-granted' practice you would like to study? Is there a new practice you

would like to study? Is there a hierarchical relationship you want to know more about? Do you want more information about a marginalized group or practice? What gap in knowledge do you want to study? Is there a binary opposite in your practice area that you want to deconstruct? What power relations do you want to explore?

2   After you have chosen an area of interest/topic, outline what you believe will be the purpose of your research. What difference will it make?

3   What is the context of your topic? What is the setting? Social, institutional, cultural, historical?

4   What do you know about the topic, from your own experience, from the literature, or from what your peers or colleagues have said?

5   Develop an argument stating why your topic is important to research – its benefits and significance.

6   Explain clearly throughout the proposal why a postmodern/poststructural approach would be useful to use. Why do you believe a postmodern/poststructural approach would be useful? How would you keep the proposed research design 'open' to explore multiple realities – to look for new, possibly suspected, or unsuspected findings. However, remember 'exploring' does not mean that anything goes. There needs to be well thought out parameters (development of research topic) before the purpose and research questions can be formulated.

Establishing the Significance and Benefits of Your Proposed Research

It is important to write a strong section on the purpose and significance of the research as this will demonstrate how what you are proposing is worthwhile. This is particularly important when using postmodern or poststructural frames in your research given the critique often levelled at these approaches as simply talking trivialities in high sounding language.[1] If using postmodern or poststructural approaches and you are applying for funding it is important the benefits to practice and/or practice settings are clearly articulated.

Depending on the intended audience of the proposal, the benefits of research might pertain to economic benefits such as reduction of costs of delivery or service provision in health care; social benefits such as increase in quality of life; policy benefits; institutional or organizational benefits; or any other type of benefit.

The following excerpt is the "benefits" section of a funding proposal using postmodern/poststructural approaches that Julianne wrote for an Australian Research Council (ARC) small grant. When reading the following "benefits" section can you determine the audience Julianne was writing for?

The benefit of this research lies in its potential to uncover the way in which Toxic Shock Syndrome has been represented in 4 examples of Australian print based media, including the often overlooked popular magazine genre, in order to explore the impact that such representation has on the construction of understandings of Toxic Shock Syndrome, tampon use, menstruation and women's health more generally. It will reveal the vested interests that may be present in the reporting of Toxic Shock Syndrome, especially in terms of whether it is a tampon related/induced health risk. Once such representation is analysed and its effects better understood it will be possible to design relevant health promotion strategies and information packages for women pertaining to Toxic Shock Syndrome, the use of tampons and other related health issues.

The benefits Julianne identified as being derived from this research fit into several categories:

- Contribution to new knowledge about the way health and illness are represented in popular print-based media—a benefit directly related to the generation of knowledge and scholarship
- Examination of competing interests in the portrayals of health and illness—a political analysis
- Practical outcomes derived from the research which can assist health promotion and the development of health information—benefits directly related to the delivery of health care and to the provision of health-related information

Thus, if Julianne had targeted an appropriate funding audience, the audience was likely to be a funding body which was interested in both the generation of new knowledge by research and practical outcomes derived from research.

It was precisely these areas which were evident in the five benefits of research that the ARC itself identified. The ARC funding guidelines asked researchers to identify one or more of the following benefits to which the proposed research relates. The five benefits stated by the ARC were:

- Contribution to the quality of our culture
- Graduates of high quality
- Direct application of research results
- Increased institutional capacity for consulting, contract research, and other services
- International links

The proposed project about Toxic Shock Syndrome relates to two of the five benefits of research specified namely:

1 Contributions to the quality of our culture in that the project enables an Australian perspective on worldwide scholarly debate about the representations of health and illness. It does so in a unique way, using sources of material that are somewhat taken for granted. The impact of popular women's magazines on particular cultural understandings in Australia (and indeed worldwide) has been relatively neglected.
2 Direct applications of research results. The project will facilitate, by analysis of the way toxic shock syndrome is represented, the design of relevant health promotion strategies and information packages pertaining to toxic shock syndrome, and the use of tampons and other related health issues for both consumers and health professionals.

Finally, it should not surprise you to find an emphasis on an Australian perspective in this research as the funding body is the Australian Research Council.

An Example of Developing a Research Proposal

Keeping in mind the preceding discussion, Julianne will now use the proposal she wrote and submitted to the ARC to explore in more detail an example of how a research proposal using postmodern and poststructural frames might be written. When doing so Julianne explores how this proposal was formed and what ways of thinking impacted on that formation. She attempts to get, as it were, under the text to expose the assumptions that this proposal as text makes in terms of how it will be understood.

Even though this proposal was written several decades ago it is still relevant as it is the principles involved in the craft of such postmodern and poststructural informed proposal writing which are the focus of our exploration. It should be possible for you to extrapolate the discussion and points made here to other grant schemes and guidelines that may vary slightly from the example here. In the discussion to follow the indented text is taken from the actual research proposal.

The title of the research project outlined in the funding proposal to the ARC Small Grants Scheme was "Constructing Toxic Shock Syndrome: Selected print-based media representations of Toxic Shock Syndrome from 1979–1995." Julianne contextualized the proposed research with a concise synopsis:

> The way in which the print based media, including popular magazines, represents health issues, influences and shapes societal attitudes towards illness and understandings of health risk. This study explores the way in which a relatively new health phenomenon, Toxic Shock Syndrome, has been represented in Australian print based media from 1979–1995. In so doing it will, by textual analysis of articles pertaining to Toxic Shock Syndrome, uncover changing and different representations of Toxic Shock Syndrome and their effect on perceptions of Toxic Shock Syndrome and related women's health issues and risks.

This synopsis outlines in broad terms, using non-technical and jargon-free language, what the proposed research is about. The synopsis is important as it is often used to locate the research in terms of which assessment panel and/or expert referees it should go to for evaluation and comment. Thus, it is absolutely critical that the synopsis and any key words that might be given convey an adequate sense of what the research is about and the approach taken so that a good match is obtained between the research proposed and the understandings and expertise of those who assess it.

Aims

Julianne wrote the following aims that clearly reflect, and are congruent with the postmodern/poststructural lens she would use in the study:

The project will

1 Locate articles relating to Toxic Shock Syndrome appearing in selected Australian print-based media between 1979 and 1995.
2 Use descriptive statistics to build a longitudinal picture of the extent, timing (month/year) and type of reporting of issues concerning Toxic Shock Syndrome in these media.
3 Analyse the content of the reporting to ascertain prevalent themes evident in such reporting.
4 Explore the discourses present about Toxic Shock Syndrome in the articles found, thereby exploring ways in which these print-based media create and sustain understandings of Toxic Shock Syndrome, menstruation and women's health.

5 Provide the basis for better understanding of the discursive construction by print based media of Toxic Shock Syndrome specifically, and women's health generally, in order to ensure appropriate
targeting of further research into representations of Toxic Shock
Syndrome and to allow for the development of more appropriate
and relevant educational programmes and information material for
women with respect to Toxic Shock Syndrome, menstruation, tampon use and other related health issues.

These aims are very clear. They tell the reader exactly what the study
hopes to achieve at each phase of the project. Further, they clearly locate
the study theoretically by referring to discourse and discursive construction. The final aim indicates what the investigator hopes to achieve in
terms of practical implications and applications for health care practice.
Once these aims have been articulated and established clearly, they guide
the development of the rest of the proposal. All material incorporated
into the proposal must clearly relate to the stated aims.

Background and Significance

In this section of the proposal, which sometimes can be called "critical
literature review," the "why" question is answered with respect to the
research being proposed. Why is this research important? Why does it
need to be done? How does it fit with other work and studies that have
been done in the area?

In other words, this section of the proposal contextualizes the research
in terms of what has gone before. It should demonstrate the researcher's
understanding of a particular area. It should also demonstrate that there
is a need for the research being proposed. Any key or core studies in the
area should be cited, including those studies which may have indicated a
need for this research to be done. It is important that the background and
significance of the project clearly relates to the aims and objectives identified in the proposal.

An Activity for You to Try to Ground This Discussion

The following excerpt is the background and significance section from the
Small ARC research proposal. It is reasonably lengthy, and we ask you to
read it with the following points in mind. Does this excerpt:

− Indicate why the proposed research is important?
− Explain how it fits with other work and studies done in the area?
− Cite key works in the area?

If so, do the cited studies:

- Indicate the need for research such as that proposed here?
- Provide information which clearly relates to the aims of the study?

You may also want to think about how this section is structured. How does it begin? How does it establish both the theoretical and substantive frames to be employed in the study? Finally, how does the discussion turn to focus on the expected outcomes and benefits of the research? Considering questions such as these will give you insights into the craft of proposal construction. You may want to make notes in the margins to assist your analysis. Such analysis can provide you with a structural frame around which you can build the background and significance sections for proposals you may write.

The background and significance section of the proposal read as follows:

> Much of the current research into health and illness asserts that many diseases are not value free, scientific entities. Instead, it is apparent that the manner in which certain diseases are represented in social texts, such as the press or magazines, is influenced by societal beliefs and values[2].
>
> (Fox, 1993; Lupton, 1994b, 1994c, 1995)

> Furthermore, such research reveals that the way diseases or health are represented may hide assumptions which underpin the very definitions of health and illness themselves. An outcome of this work is that understandings of health, illness, and disease processes cannot be viewed as independent of the social context in which they are situated.
>
> Lupton (1995) asserts that the news media is one of the most important contemporary forums for the articulation of discourses, that is, particular ways of thinking and talking about reality (Foucault, 1980), relating to health and medicine. "A discourse provides a set of possible statements about a given area, and organises and gives structure to the manner in which a particular topic, object, process is to be talked about" (Kress, 1985, p. 7). Who is quoted, who is not; what sources are used and how an article is framed affects perceptions of the health issue being reported. News is thus a presentation of a particular view of reality, not a value free reflection of facts. "All news is always reported from some particular angle."
>
> (Fowler, 1991, p. 10)

> For example, Gabe et al. (1991) analysed the content of newspaper coverage on the issue of women's tranquilliser dependence. Their

analysis highlighted the importance of taking into account ways in which mass communication can distort and bias understandings about dependence, how it can reaffirm stereotypical understandings about women's passivity, and how it can thereby conceal crucial difference between groups or individuals who are dependent on tranquillisers. Chrisler and Levy (1990) found, on analysis of media content on Premenstrual Syndrome (PMS), that the messages conveyed by the media confused biologically based premenstrual changes with the actual syndrome, so that PMS became so inclusive in focus that it would be difficult for women not to find at least part of their experience recorded. Further, Stallings (1990) has demonstrated that perception of health risk is not so much the outcome of media and public discourse as existing in and through processes of that discourse. "Hence risk is never constant. It is created and recreated in discussion of events that are seen to undermine a world taken for granted."

(p. 82)

Likewise, articles and reporting in popular magazines, a largely overlooked area, reflect certain representations of reality. Baird and Sheridan's (1992) analysis of the Australian Women's Weekly leads them to assert with respect to developing an index for the popular women's magazine, "the most interesting feature of the construction of the index thus far has been the insight into the reader positions that the Weekly constructs."

(p. 150)

Albert's (1986) study of how a sample of nationally circulated magazines in the US treated and represented a new medical phenomenon—AIDS—clearly demonstrated the effect of this genre of print based media in creating certain understandings and perceptions of AIDS.

Studies such as these alert researchers to the need to consider the role of media representations themselves in constructing images of health/illness, and the social understandings of health/illness that result from such representations. As Nelkin (1991) asserts, newspapers and popular magazines:

filled with health advisory columns, as well as news about risk events, dominate the avenues of public information...The information they convey, their visual and verbal images, and tone of the presentation can define the significance of events, shape public attitudes and legitimate or call into question public policies.

(p. 302)

This is true of the reporting about Toxic Shock Syndrome.

Toxic Shock Syndrome (TSS) was first named as such in 1978 by James K. Todd (Hanrahan, 1994) in Denver, USA. However, it possibly has been present in the medical literature since 1927 usually described as staphylococcal scarlet fever (Colbry, 1992). Toxic Shock Syndrome symptoms include the sudden onset of high fever, vomiting, shock, body rash and peeling skin. Toxic Shock Syndrome's specific causative agent is the bacterium staphylococcus aureus. Whilst the majority of Toxic Shock Syndrome patients are menstruating women; non-menstruating women, children and men have contracted Toxic Shock Syndrome. Toxic Shock Syndrome has been linked to tampon use during menstruation but the exact nature of the linkage between tampons, the bacterium and Toxic Shock Syndrome remains problematic. The link between Toxic Shock Syndrome and tampon use was first reported in mid 1979 and early 1980 when Toxic Shock Syndrome resulted in the hospitalisation of 7 people in Wisconsin, USA—6 of whom were menstruating women using high absorbency tampons and one of whom was male (Olesen, 1986). This led to massive media coverage and the subsequent construction of Toxic Shock Syndrome as an emergent new health risk. How the media reported on, and thus constructed understandings of Toxic Shock Syndrome from its discovery in 1978 up until the present (end of 1995) is the focus of this study.

Discourses surrounding Toxic Shock Syndrome are particularly problematic as Toxic Shock Syndrome touches on taboo areas such as menstruation and sanitary products.

(Delaney, Lupton & Toth 1988)

Further, there are competing and vested interests in portrayals of Toxic Shock Syndrome such as those of tampon manufacturers, health authorities, consumer groups and so on. As Olesen (1986) declares:

> the toxic shock phenomenon poses critical questions in the definition and construction of the issues... certainly in the case of toxic shock syndrome, different definitions, predicated on the production of research data and the presumed confounding influences of the mass media of communication were and remain in play.
>
> (pp. 57–58)

Indeed, in a recent edition of the Australian Medical Journal, Garland and Peel (1995) declare Australians have been subjected to misleading information about Toxic Shock Syndrome in media publicity about the recent death of a girl in Queensland.

This study aims to address the question of the way in which Toxic Shock Syndrome has been defined and constructed as a health issue, since its formal recognition in 1978 and link to tampon use in 1979.

It will provide a systematic overview of some of the Australian print based media reporting that has been done about Toxic Shock Syndrome in order to explore how Toxic Shock Syndrome has been constructed and what interests have been represented in that construction. Such analysis is essential if effects of the reporting, especially with respect to Toxic Shock Syndrome as a public health issue, are to be exposed and understood. Only then can moves be made to reveal Toxic Shock Syndrome from all angles—not just those constructed by certain media. This is imperative as "the media provide and reflect blueprints for action" (Clarke, 1991, p. 304). The significance of the study thus lies in attempting to answer (at least in part) Clarke's question "if the media reflect biased views of disease, its causes, meanings and best treatment alternatives, what can be done to ensure more appropriate portrayals?" (p. 305). Timely and responsible reporting is critically important to the management of Toxic Shock Syndrome (Colbry, 1992).

How Did Your Analysis and Exploration of the Way in Which the Background and Significance Sections Have Been Crafted Go?

When reading this background and significance section of the proposal, were you able to identify the structural framework underpinning them? Points to note about such a framework include (and you may find it useful to mark the sections of the background and significance section that pertain to each of the points made in order to get a "visual" sense of how this section was constructed):

- The section begins at a macro theoretical level. It discusses textual representations of disease and illness and posits that any such representation conceals assumptions and understandings about the definitions of health and illness
- Having established the broad theoretical frame, the discussion moves "down" one theoretical level to look at the way two types of texts—news media and popular magazines—have represented aspects of health care practice and health care reality. Studies are cited for each type of text, and these studies are used to demonstrate the possibilities for analysis of health issues afforded by exploring the way these texts portray and represent disease and illness
- This opening part, which contextualizes the research theoretically and in terms of studies that have been done in the area, ends by asserting the importance of the type of study proposed, arguing from Nelkin (1991) that such representation can "define the significance of events, shape public attitudes and legitimate—or call into question—public policies" (p. 302)

- The discussion then turns to look at the particular health topic that will form the substantive focus of the proposed research. After describing Toxic Shock Syndrome (TSS), the research is brought into clear focus at the end of the introductory discussion about TSS with the statement: "How the media reported on, and thus constructed understandings of Toxic Shock Syndrome from its discovery in 1978 up until the present (end of 1995) is the focus of this study." This statement relates clearly to the stated aims of the project. Thus, there is already congruence between the various sections of the proposal. Such congruence is an essential part of the craft of proposal writing
- The discussion then turns to introduce the notion of discourses that shape the way TSS is written, spoken and thought about
- In terms of setting the scene for the background and significance of the study, the discussion returns to the aims of the study and discusses how these fit with the concepts and ideas that have arisen in the background section
- The last point in this section reinforces the significance of the study and relates it to health care practice.

In summary, the background and significance sections locate the study in a theoretical sense, where such "location" may operate at various levels of theoretical analysis. It then links the theoretical frame of the study to the substantive area under discussion and demonstrates how the theoretical approach taken can lead to new and important insights into the area under discussion. In so doing, it refers to other studies that can illuminate aspects of the proposed research. Finally, the discussion clearly articulates the aims of the study. It emphasizes the significance of the study in terms of its implications for, and application to, the field of research (in this case health care delivery, practice, and understandings).

Research Plan, Methods, and Techniques

Having established the "why" and "what" in relation to the proposed research, a proposal will then need to outline the "how." How will the research be carried out? It is important that the research plan be congruent with the theoretical orientation of the study and with its stated aims. One of the issues that confronts researchers when writing this section of the proposal is how much detail to include about the methods themselves and the actual analytic techniques. One issue to consider is whether some terms could reasonably be expected to be understood by reviewers[3] – in particular, terms related to postmodern and poststructural thinking. For example, terms such as discourse, discourse analysis, and deconstruction. You will need to give clear explanations of terms and ideas that may be unfamiliar to the possible reviewers and/or review panel of this scheme. If the terms *are* familiar to reviewers, they will still want to see that you also understand them as this will enable them to feel confident that you know what you are talking about and be able to implement the study.

We will demonstrate how such clarity might be achieved using the research plan, methods, and techniques section of the TSS grant proposal. As you read this excerpt, ask yourself the following questions:

– Does the methodology "fit" with the theoretical orientation of the study?
– Is there a clear link with the aims of the project?
– Will the research plan outlined enable the researcher to achieve the aims?
– Is the proposed research feasible?

Excerpt: The Research Plan, Methods, and Techniques Section of the TSS Proposal

The study comprises an exploratory critical analysis (see Lupton, 1994b) of the reporting of issues associated with Toxic Shock Syndrome in selected Australian print media from 1979-1995. In so doing it incorporates traditional content analysis studies of media (Bell et al., 1982; Winston, 1990). However, the focus does not rest only on the manifest (Lupton, 1994b) or surface content as many traditional content analyses do.

Instead the study seeks to probe the latent or subtextual (Lupton, 1994b) discourses (Foucault, 1977) that both construct and are constructed, about Toxic Shock Syndrome.

The methodology section draws on concepts presented in the background/significance section of the proposal. The methods section goes on to give specific detail as to how this will actually be done.

Methods

Working within the framework of critical analysis, the study will survey and locate all articles pertaining to Toxic Shock Syndrome appearing in 4 purposively selected (after Clarke 1991) print based Australian media from 1979-end of 1995. 1979 has been chosen as a starting point for the study as Todd identified Toxic Shock Syndrome in late 1978. The sample is purposive in that the 4 print based media have been chosen to reflect diverse Australian target audiences thus providing a potentially rich supply of data in terms of the angle from which each reports and constructs Toxic Shock Syndrome.

The print based media to be surveyed comprise:

1 The Australian Women's Weekly
   A once weekly, now monthly, women's popular magazine with a circulation (1995) of 1 million and estimated readership of 3.4 million, chosen because of its long standing place in Australian culture as a women's magazine of broad appeal. As McNicoll (1982, p. 40) puts it, "almost everyone at some time or other has read it...its appeal crossing the barriers of the sexes, a magazine aimed at women but finishing up with a unisex appeal."

The same sort of detail was included for three other print-based media namely Cleo, Dolly and The Advertiser but have not been included here for space reasons.[4]

Method of Locating Articles

1 The Australian Women's Weekly
   This magazine is not indexed, so a manual search is the best option for locating articles. Back issues are held by the Flinders University Main Library on microfilm from the first issue (June 1933) up until 1983. The South Australian State Library (Mortlock) holds issues on microfilm until December 1994, and hard copies for 1995 (active subscription). Photocopies may be obtained of any articles. Photocopying is cheaper from Flinders University so articles up until 1983 will be obtained from there, and subsequently from Mortlock Library.

The same sort of detail was included for Cleo, Dolly and The Advertiser but have not been included here for space reasons.[5]

Tips for deciding what data will be collected in a study informed by postmodern thinking

Here are some questions that will help you decide how and what information you will need to collect. What type of 'text' or data will you need to collect to answer your research question? For example: participants lived experience? Spoken or written? Journals or interviews? As you think about and choose how to answer these questions, you will need to think about how your choice(s) will help you answer your research question. The level of detail provided in the above excerpts from the TSS study proposal clearly demonstrates the extent to which the researcher has thought the research process through. This is very important. The research plan must be precise, concise, and above all, manageable. There is a somewhat deceptive simplicity in good research design.

Analysis of the Articles

The next part of the proposal needed to outline how the data would be analysed. Here is what the proposal said.

> Having obtained the articles a triangulation (Denzin, 1989) of methods will be used in the analysis in order to provide a "thick description" (Denzin, 1989). Such triangulation involves 3 phases of analysis:
>
> **Phase 1**. Each article obtained will be analysed in terms of:
>
> a) Headline
> b) Topic
> c) Visual material
> d) Sources quoted
> e) Whether it is primarily theoretical or practical
> f) Tone—positive or negative?
> g) Position in paper/magazine
> h) Relationship (if any) to other news articles at the time
> i) Type of print media it appears in
> j) Type of audience article pitched at
> k) Day of week
> l) Month of year
> m) Year
>
> Descriptive statistics will be applied to the above categories in order to build a picture of the extent, range, location, timing and type of reporting of issues concerning Toxic Shock Syndrome in these media. This draws on categories of analysis from studies by Lupton (1994a)— categories a, b, c & d; Gabe, Gustaffason and Bury (1991)—categories i & j; and Chrisler and Levy (1990) categories e and f Additional categories (g, h, k; l & m) will further describe the construction of Toxic Shock Syndrome as a health issue from 1979-1995.
>
> **Phase 2**. Qualitative thematic analysis of the articles will be employed in this phase.
>
> The actual text of each article will be thematically coded then analysed. Both manifest and latent content will be themed.
>
> **Phase 3**. The data obtained from Phase 1 and Phase 2 will be used to explore the discourses present about Toxic Shock Syndrome in order to develop an analysis of the representation of this health issue. This paves the way for further studies such as the effect on women of the way Toxic Shock Syndrome is represented in the media; and their understandings of Toxic Shock Syndrome, menstruation, tampon use, and other related health issues. Further, an interesting study could be done on Australian media representations of Toxic Shock

Syndrome as opposed to, say, those in the US. A limitation of this study is that is restricted to 4 print based media only. A larger study could analyse more print based media sources.

You can find a detailed description of how this analysis was conducted and what it revealed in the published article about the study.

Cheek, J. (1997) (Con)textualizing Toxic Shock Syndrome: Selected Media Representations of an Emergent Health Phenomenon 1979-1995, *Health, 1*(2), pp. 183–203.

Summing Up Where We Are

Having read these excerpts about the research plan, methods and techniques, and analysis, let us return to the questions posed at the outset of this section, page 85, and consider each in turn.

1 **Does the methodology "fit" with the theoretical orientation of the study?**
  The theoretical orientation of the study problematizes the representation of health and illness by print-based media, arguing that such representations reflect certain understandings of health/illness and conceal others. Thus, the methodology employed should enable the way representations of health/illness are discursively constructed to be explored. As an exploratory discourse analysis the study enables such an explanation.
2 **Is there a clear link with the aims of the project?** Yes, there is! In fact, if you look closely at the structure of the "methods" section you will notice that it is framed around the aims of the study. For example, the section clearly outlines what articles will be located and how this will be done (see aim 1 in the earlier section "Aims"). It also contains a clear statement about why these particular media were chosen (a purposive sample) which is very important in terms of the research design. A substantial amount of detail is given about the methods for locating the articles as it is important to establish the feasibility of the study, whether the articles can in fact be obtained, and what are the likely costs associated with locating them.
3 **Will the research plan outlined enable the researcher to achieve the aims?**
  The methods section then moves on to address how each article will be analysed: descriptively in Phase 1 of the analysis and thematically in Phase 2. This directly relates to Aims 2 and 3 of the study. Finally, Phase 3 of the research plan outlines the discursive phase of the analysis, which relates to Aim 4. Thus, not only have clear links been made to the aims of the study throughout the various sections of the proposal; the aims have in fact provided the framework around which the discussion in the

methods and analysis sections of the proposal have been structured. As there is congruence between the aims and the proposed methods and analysis, the research plan should enable the aims to be met.

4 **Is the proposed research feasible?** The feasibility of the proposal was addressed in the detail provided about the location of each magazine/ newspaper and in the way in which articles would be accessed. This is important, as otherwise it may have been difficult to assess whether what was being proposed could in fact be done!

To sum up, the proposal clearly situates this research in terms of other studies in this substantive area. Having outlined the overarching methodological frame, it went on to outline specific methods for collecting the research materials, including a rationale for the four print-based media chosen for the study. At all times there was a clear link between the methods being proposed and the aims of the study.

Having established the mechanics of collecting the articles, the proposal outlined the way in which the articles would be analysed. The source of the research material should be very clear to the reader of a proposal as should the ways in which it is going to be analysed.

In her book Mixed Methods Research: merging theory with practice Sharlene Hesse-Biber (2010) described this study as an example of a postmodern inspired study using mixed methods comprised of 3 phases. These were 1. Quantitative analysis and interpretation 2. Qualitative analysis and interpretation 3. Qualitative analysis and interpretation, with the findings of each stage being integrated to compare and contrast findings with the goal of getting at dominant and hidden discourses in media portrayal of toxic shock syndrome.

You can read more about Hesse-Biber's (2010) analysis of the methodology Julianne employs in Hesse-Biber's chapter on Postmodernist approaches to mixed methods (pp. 154–173) which also has other examples of studies using postmodern approaches and mixed methods.

Timetable and Budget

Other sections of the proposal to be addressed are the time frame/time line for the research and the budget. The latter should include a justification for each of the budget items requested, while both the budget and the time frame should relate clearly to the rest of the proposal. For example, the time frame for the grant about Toxic Shock Syndrome read as follows:

Timetable

Months from Receipt of Grant

1-4: Phase 1
Location and collection of articles (see aim 1). Progressive entry of descriptive statistics as per Phase 1 in Research Plan (see aim 2).
Completion of phase 1
5-9: Phase 2
Qualitative thematic analysis of articles and coding as per Phase 2 in Research Plan (see aim 3).
Completion phase 2
10-12: Phase 3
Exploration of discursive construction of Toxic Shock Syndrome and analysis of possible effects on women's understandings of Toxic Shock Syndrome and associated health issues as per Phase 3 in Research Plan (see aim 4).
Completion and production of final report (see aim 5).
Report submitted to ARC, University of South Australia and any other interested parties.

Here the proposed sequence and timing of the research clearly aligns with the phases outlined in the research plan and the aims of the project.

Once the way that the project was to be done was completed, it was then possible to 1. Identify what costs would be involved 2. Provide justification for each of those costs. The justification of each component of the budget must clearly relate to the identified phases and aims of the project. This is very important as otherwise the reason for requesting a certain component of the budget may not be clear. For example, if you are requesting research assistance, it is important to state clearly in detail why you need the assistance and how it pertains to the research plan and aims of the project.

Ethical Considerations

This study did not require formal ethics approval such as from an ethics committee or Institutional Review Board (IRB) as it was using published texts from the public domain. However, there were still ethical considerations that needed to be considered and acknowledged. These included that the research was feasible (i.e. could be done in the way proposed and according to the timelines and budget), and that the research was clear in terms of its purpose and its questions/areas of inquiry were focused enough so that the research was likely to be able to achieve its stated outcomes. Also, that the researchers had the skills and knowledge to carry out the research.

When writing proposals for research considerations regarding the ethics of any piece of research, including research from postmodern and poststructural approaches, such considerations include:

- Clear statement of the issue/problem
- Background and literature review to support purpose of research from an ethical point of view
- Purpose of the research and research question
- Confidentiality of participants and data
- Safety of participants
- Value as opposed to risks of the research
- Data handling and storage
- Informed consent of participants
- Right of participants to withdraw from the study at any time

As Cheek and Øby note (2023) "there is much more to research ethics and ethics committees than regulation and procedures...(E)thics, and thinking about ethics impacts on all aspects of designing research" (pp. 41–43).

An accessible and reflexive discussion "of research ethics as an orienting idea framing and permeating the entire research process" (Cheek & Øby, 2023, p. 27) is provided by Cheek and Øby in Chapter 2 of their book Research Design. Why Thinking About Design Matters. It raises important points to be taken into account when proposing research – including research using postmodern and poststructural approaches.

Concluding Comments

This chapter has been about the craft of proposal writing and how this craft affects the development of research proposals drawing on postmodern and poststructural perspectives. The discussion has been grounded in a research proposal written by Julianne and analysed as text to illuminate principles involved in the construction of a proposal generally, and more specifically a proposal that draws on postmodern and poststructural thought. In particular, the need for congruence between the aims, the methods and the theoretical orientation of the project has been emphasized.

We continue our discussion of proposing research using postmodern and poststructural thinking in the next chapter. We focus on how to deal with, and respond to reviewer comments and critiques as part of the process of seeking funding and/or approval for your research.

Notes

1 See discussion of these critiques in earlier sections of the book – for example Chapters 2 and 3.
2 The references were the most up to date relevant ones related to the study at the time of writing the proposal.
3 We pick up on this point in Chapter 7.
4 You can find that detail in the first edition of this book Cheek J (2000) Postmodern and Post structural Approaches to Nursing Research, Sage Publications, p. 84.
5 You can find that detail in the first edition of this book Cheek J (2000) Postmodern and Post structural Approaches to Nursing Research, Sage Publications, p. 85.

References

Albert, E. (1986). Acquired immune deficiency syndrome: The victim and the press. *Studies in Communication, 3*, 135–158.

Baird, B., & Sheridan, S. (1992). Indexing the Australian Women's Weekly, *The Australian Library Journal*, May, pp. 145–150.

Bell, P., Boehringer, K., & Crofts, S. (1982). *Programmed Politics: A Study of Australian Television*. Sydney, Sable.

Cheek, J., & Øby, E. (2023). *Research Design. Why Thinking about Design Matters*. SAGE.

Chrisler, J., & Levy, K. (1990). The media constructs a menstrual monster: A content analysis of PMS articles in popular press. *Women and Health, 16*(2), 89–104.

Clarke, J.N. (1991). Media portrayal of disease from the medical, political, economy, and lifestyle perspectives. *Qualitative Health Research, 1*(3), 287–308.

Colbry, S.L. (1992). A review of Toxic Shock Syndrome: The need for education still exists. *Nurse Practitioner, 17*(9), 39–46.

Delaney, J., Lupton, M.J., & Toth, E. (1988). *The Curse: A Cultural History of Menstruation*. Champaign, IL, University of Illinois Press.

Denzin, N. (1989). *The Research Act*. Chicago, Aldine.

Foucault, M. (1977). *Discipline and Punish*. London, Tavistock.

Foucault, M. (1980). *Language, Counter-Memory, Practice: Selected Essays and Interviews*. D. Bouchard (Ed.), Ithaca, New York, Cornell University Press.

Fowler, R. (1991). *Language in the News: Discourse and Ideology*. London, Routledge.

Fox, N. (1993). *Postmodernism, Sociology and Health*. Toronto, University of Toronto Press.

Gabe, J., Gustaffason, U., & Bury, M. (1991). Mediating illness: Newspaper coverage of Tranquilliser dependence. *Sociology of Health and Illness, 13*, 332–353.

Garland, S.M., & Peel, M.M. (1995). Tampons and toxic shock syndrome. *The Medical Journal of Australia, 163*, 8–9.

Hanrahan, S.N. (1994). Historical review of menstrual toxic shock syndrome. *Women & Health, 21*(2/3), 141–165.

Hesse-Biber, S.N. (2010). *Mixed Methods Research: Merging Theory with Practice*. Guilford Press.

Kress, G. (1985). *Linguistic Processes in Socio-cultural Practice*. Victoria, Deakin University Press.

Lupton, D. (1994a). Analysing news coverage. In S. Chapman & D. Lupton (Eds.), *The Fight for Public Health: Principles and Practice of Media Advocacy*, London, BMJ, pp. 23–57.

Lupton, D. (1994b). Femininity, responsibility, and the technological imperative: Discourses on Breast Cancer in the Australian Press. *International Journal of Health Services*, *24*(1), 73–89.

Lupton, D. (1994c). *Medicine as Culture: Illness, Disease and the Body in Western Societies*. London, Sage.

Lupton, D. (1995). Anatomy of an 'Epidemic': Press reporting of an outbreak of Legionnaire's Disease. *Media Information Australia*, *76*(May), 92–99.

McNicoll, D. (1982). The Weekly: A 50-year Phenomenon. *The Bulletin*, Oct 12, pp. 40–48.

Nelkin, D. (1991). AIDS and the News Media. *The Milbank Quarterly*, *69*(2), 293–307.

Olesen, V.L. (1986). Analyzing emergent issues in Women's Health: The case of toxic-shock syndrome. In V.L. Olesen & N.F. Woods (Eds.), *Culture, Society, and Menstruation*, New York, Hemisphere Publishing Corporation, pp. 51–62.

Stallings, R. (1990). Media discourse and the social construction of risk. *Social Problems*, *37*(1), 80–95.

Winston, B. (1990). On counting the wrong things. In M. Alvardo & J.B. Thompson (Eds.), *The Media Reader*, London, British Film Institute, pp. 50–64.

7 Navigating Challenges when Proposing and Conducting Postmodern/Poststructural Research Activities

In this chapter we will

- Develop the idea introduced in the previous chapter that proposals are texts crafted for particular audiences
- Highlight that there are written and unwritten "rules" for the form that research proposals take
- Present ways to think about navigating these "rules" when writing a research proposal using postmodern and poststructural approaches
- Suggest strategies to respond to reviewers' comments about a proposal using postmodern and poststructural approaches'

Introduction: Research Proposals as Texts Full of Assumptions

Research proposals are examples of text. Like other texts such as journal articles, books, theses, and case notes[1] they take a very specific form. This is because all research proposals are written for particular audiences. These audiences include funding bodies, ethics committees and, if the proposal forms part of a student's programme of study, course review committees. Each audience has its own expectations—its own written and unwritten rules for how the proposal should be framed. These expectations are shaped by beliefs (often unwritten) about what is, and thus what is not, appropriate subject matter for research, as well as what is, and thus what is not, a suitable way of carrying out the research.

For example, if a particular course review committee or funding body's instructions/guidelines for the presentation of a research proposal insist that a hypothesis be given, then the unwritten assumption in operation is that all research should conform to the form demanded by traditional notions of what a scientific method is. It effectively excludes other types of research for which hypotheses are not relevant. The researcher who wishes

DOI: 10.4324/9780429053764-7

to use research approaches for which hypotheses are not appropriate is left in a situation in which their research does not "fit" the assumed form that a research proposal should take. For example, a discourse analysis or a deconstruction study will not conform to such assumptions, nor will research that is designed to be exploratory and interpretive.

It is important to recognize that this does not mean that research which does not "fit" such assumptions is alternative and somehow less valuable. Rather, what it does mean is that such research does not conform to one set of understandings about what research might be, what it might involve, and how knowledge develops. There are some dissertation committees and funding bodies that have responded to this problem and use the term 'research question' instead of hypothesis, to ensure inclusivity of, and applicability of their guidelines, to both quantitative and qualitative approaches. However, the language of hypothesis driven research continues to be the norm rather than the exception in many instances. It also continues to be the norm in terms of framing the way that many course review committees or funding bodies (and the reviewers they use) evaluate the research proposals submitted to them (Cheek, 2022).

Along with such unwritten rules are written rules in the guidelines for filling in research application forms which specify the format an application must take. These may include what should be discussed, how much is required to be discussed (i.e., length of each section), and by default, what does not need to be discussed. These rules "force" researchers to adopt the specified format. To not do so would inevitably result in the application being assessed as incomplete and consequently not able to be funded. It can be very difficult to "force" postmodern/poststructural inspired research into such formats. How much "forcing" is needed can therefore be another good guide to the assumptions held by a funding body about the form research should/must take and whether they are open to research informed by postmodern/poststructural thought.

It is also important to consider whether the outcomes of the research will be of interest to the proposal's audience. For example, a course review team might consider a theoretical treatise on some aspect of health care practices a perfectly reasonable research outcome whereas a funding body concerned with developing cost-efficient modes of health care delivery may be less likely to see the value in that theoretical work.

Research proposal writing is, at least to some extent, a political process (Cheek, 2022). As we discussed in Chapter 6 all proposals for research are written for particular audiences. If a postmodern or poststructural approach is to form part of a proposed piece of research, it is important to ascertain that the audience who will be reviewing the proposal is amenable or open to considering that research approach. This is particularly so with respect to funding bodies. Some funding bodies will only fund certain types of research, whereas others may be more open to a range of approaches.

A tip

It is important to be aware of the expectations and assumptions (both spoken and unspoken) of the audience for whom you are writing. Consider if the approach to research you wish to take and the outcomes you are proposing fit with the agenda of that audience. If they do not, then it is unlikely that you will be successful in your application. Search for an audience that seems amenable to the approach you are suggesting. There is nothing to stop you from writing to course review committees or funding bodies that seem closed to the approaches you are suggesting and pointing out the difficulty that their application guidelines/instructions pose for certain types of research. Our own experiences have demonstrated that sometimes those assessing research do not realize how restrictive the guidelines they have developed are for researchers. It is possible to agitate for change. Julianne found herself invited to address funding panels to discuss the issues she had raised in a letter, and in one instance ended up assisting in the revision of the application form and guidelines so that they became more inclusive of different approaches to research. Researching the audience for whom you are writing is thus, in some ways, of as much importance as the development of the actual research proposal itself!

In the next section we will explore the effect of these unwritten and written assumptions, and how they can be navigated, by exploring the review of a proposal that Megan developed. Although eventually successful, this proposal had to be revised and resubmitted several times before being funded. We will take a look at the reviews that led to those revisions, and what assumptions on the part of the reviewers they were based on. We will also share and discuss how Megan responded to those reviewers' remarks thereby giving unique insights into another part of the proposal development process that often is not made visible in most research.

Navigating Challenges When Writing and Defending a Proposal Seeking Funding for Research Using Poststructural Approaches

Writing and submitting a research grant proposal for funding can be a daunting task. It is the exception rather than the rule to obtain funding since most National Funding organizations have very low funding rates – for example in Canada at the time of writing the book the average rate of funding success is 15%–20%.[2] Further, more quantitative research proposals are funded compared to qualitative studies.

In many funding schemes historically, and even today, there are more reviewers who are quantitatively trained as opposed to those trained in qualitative research (Cheek, 2022, 2024). This is problematic as often when quantitative reviewers assess qualitative research proposals the comments are indicative of the quantitatively trained reviewer not fully comprehending qualitative research. Such lack of expertise in qualitative research is often reflected in a lower score for qualitative projects which are read and assessed using a quantitative lens.

In Canada, the Canadian Institute for Health Research now requires reviewers to declare if they are able to review certain proposal applications based on their area of expertise. This is one example of how funding agencies are responding to the issue of reviewers without expertise in the approaches they are reviewing, doing those reviews when they do not have appropriate expertise to do so.

Further complicating this is that postmodern/poststructural research studies use complex concepts that need to be clearly articulated and not all researchers and reviewers, even those who are experts in other areas of qualitative inquiry, will be familiar with postmodern and poststructural approaches. Therefore, it is essential that you are aware of this and clearly demonstrate how you will incorporate the thinking associated with these approaches throughout your entire proposal.

For example, you need to 1) Create the argument for why a postmodern/poststructural approach is a suitable theory to use 2) Use a postmodern/poststructural lens to critique the literature you provide in your background/literature review 3) Write a clear purpose for the research and a research question that connects to postmodern/poststructural concepts 4) Write a comprehensive description of the concepts and finally 5) Include specific details regarding how you will use the concepts throughout your study including the way you think about ethics, data collection, analysis, and knowledge translation. The research will then be sent to reviewers for comment. You need to make sure that 1–5 above are clear and able to be understood even by reviewers who may not be experts in these approaches. The reviews will soon tell you whether you achieved this, and what you might need to do further in this regard.

In the rest of the chapter we focus on how to respond to reviews and the critique in them. We use the example of the reviews received for one of Megan's grant proposals that required three iterations before being funded. By this we mean the proposal was submitted (P1), reviewed, and rejected, then reworked in light of the reviews (P2), submitted, reviewed and rejected again. After this, the proposal was reworked yet again a third time (P3), reviewed, and this time won funding.

Megan explores how the team navigated the challenges of this iterative process of submission-review-resubmission. This exploration provides insights for how readers might navigate such a review process – be it for seeking funding, and/or meeting the requirements of a course review committee or any other audience where they are submitting their proposed research. Excerpts of the teams' responses to reviewers are italicized throughout the discussion to follow.

To better contextualize the discussion to follow you can read about the study in published reports of the research (Aston et al., 2014a, 2014b; Breau et al., 2016 – see the full reference details for these articles at the end of this chapter). You can also view a video based on the study's findings. **https://vimeo.com/163234949**.

Megan's Story: Navigating the Challenges of the Process of Proposal Development: Submission+Review+Redevelopment+Resubmission...

Our research was designed to examine the hospital experiences of children with intellectual disabilities (ID), their parents, and nurses who cared for them using feminist poststructuralism (FPS) an approach which combined poststructural and feminist thoughts. We knew the funding agency we were submitting our grant proposal to received very few studies using poststructural approaches. Although we had to submit our proposal three times, before getting funding we were determined to keep going until we got favourable reviews because we believed our proposed research was sound.

Sharing the challenges we encountered when responding to reviewers' comments will help readers in the same situation. When doing so we have chosen to focus on, and share, the second set of reviewers' comments (i.e. the comments on P2). This was because they were similar to the first set of reviewers' comments (i.e. the comments on P1). Another reason is that we want to show how the second set of comments were addressed in a way that satisfied reviewers and led P3 to winning funding.

Similar to Julianne's research proposal described in Chapter 6 we began our proposal with a clearly articulated introduction, purpose, and statement about the significance of the research in a way that captured the reader's attention. This focus was then carried throughout the entire proposal. For example, in our study we first identified the problem as *'children with ID spend more time in the hospital and experience more stigma and marginalization than typically developing children.'* We then supported this claim with literature, and demonstrated that very little literature existed that examined the experiences of children with ID, their parents, and nurses who cared for them, and no research studies using poststructural or postmodern approaches had been conducted in this area. We argued that it was imperative to include the concepts of relations of power, agency, and subjectivity to thoroughly understand experiences of participants and the social and institutional constructs in which they participated.

To address this problem, we proposed to collect the views of children with ID, their parents, and nurses who worked with them about their experiences in hospital. We wanted our research to contribute to improved communication and, ultimately, improved care so that negative effects of hospitalization were minimized. We also began to weave in the language of feminist poststructuralism by stating the following:

> *The use of feminist poststructuralism and discourse analysis will help to uncover the dynamics of different and similar beliefs, values, and meanings that children, parents, and nurses have about complex care that is physical, emotional, and social. These differences and similarities ultimately are played out through relations of power. A complex understanding of these dynamics will go beyond an individualistic and behavioural perspective and include a focus on personal, social, and institutional aspects that will add to a more complex understanding of experiences.*

This then led nicely into our research questions:

Research Questions: *How do children with ID, their parents, and nurses experience care while interacting with each other during the child's time in hospital? How are experiences socially and institutionally constructed through relations of power?*

For both the first (P1) and second (P2) submissions the reviews were blinded so we did not know the names of the reviewers. However, we were quite sure that there was at least one quantitatively trained reviewer for both submissions (P1 and P2) who was judging these submissions using a quantitative lens. This was because the questions posed by that reviewer arose from reading the proposal using quantitative understandings. For example they asked us to justify why we had a small sample size when our sample size was 60 – qualitative researchers would perhaps suggest that a sample size of 60 was not small and in fact overly ambitious.

> You will remember reading in Chapter 4 our discussion of how the review process is influenced by dominant discourses about what research is and what it is not. Specifically, Koro-Ljungberg et al. (2015), wrote about how assumptions regarding 'good science' and 'not acceptable science' could significantly impact the review process. This is exactly what we encountered.

Despite this critique we received relatively high scores on our first (P1) and second (P2) submissions, but not enough to be funded. Reviewers' comments regarding the background, literature review, and purpose of study in both submissions were positive. However, reviewers clearly did not understand the concepts within feminist and poststructural approaches. We realized that we needed to add more clarity to this section as well as specific details from a poststructural lens. Therefore, in the third submission (P3) of the application the team provided even more detail regarding theory and methodology (see 'methodology' section below). The reviewers were very positive about this addition and awarded scores sufficient for the grant to be funded.

In the discussion to follow you will see how each of the reviewers' comments about P2 were addressed. You will also see that at times this meant either agreeing or disagreeing with comments made *and* providing a clear rationale for why.

Responses to Reviewers

The Canadian Institute for Health Research which was the funding body we were applying to give applicants the opportunity to respond to the comments and critiques from reviewers about the proposal. In the discussion to follow, we analyse point by point the reviewers' comments and critique about proposal P2. We then include the actual response (*in italics*) that we sent back to the funding body about the comments. You will see that our comments serve two purposes. The first is to address the critique and the second is to provide explanations about tenets of qualitative inquiry.

Responses to Reviewer 1 about P2:

- This reviewer had a more positivistic point of view of research reflected by their suggestion that we should 'use a model to guide our research study.' We were surprised by this comment as the intent of our study was to 'explore' the experiences of participants rather than have their experiences fit into an already existing model.

A poststructural approach would question the status quo and look for undiscovered experiences. We needed to respond to this comment and make it clear that our theory and methodology would open up examination of our research issue instead of controlling and limiting our findings. We wrote

> *'This reviewer questions whether a theoretical model should be used to guide this project. Because so little research exists regarding these issues, we feel that use of a model would be premature and could constrain the richness of data that we could obtain with the exploratory approach we describe.'*

- Reviewer 1 also suggested we needed to collect information regarding past hospitalizations of the children to cross check the validity or the truth about what they were saying. We disagreed with this suggestion since feminist poststructuralism supports the concept that participants should tell their own story and their experience should not be checked or triangulated through other means. We wrote the following.

> *'We have added a brief demographic questionnaire for children and parents to complete. We expect detailed information regarding the child's previous hospitalizations to surface during the interview. The demographic questions are intended solely to group children for cross-case analyses and to set a backdrop for the interview to proceed. Because this study aims to understand the perceptions of children and parents regarding hospitalization, their recall, regardless of accuracy in relation to hospital records, is more germane to the purpose of the study.'*

- The same reviewer was also concerned about 'checking' or 'monitoring' the ability of parents to tell us the mental age of their ID child as they doubted the ability of parents to be able to identify the mental age of their child. The reviewer thought the information should be checked by medical experts. Poststructural approaches enable researchers to focus on participants' beliefs, values, and practices and challenge normative discourses, in this case, we wanted to challenge a medical discourse that used surveillance and control (see Chapter 3). In order for participant subjectivity and agency to come forth, we decided at the very least, we should respect their point of view and **not** check it with hospital records. We wrote:

> *'The reviewer asks if we will consult medical records regarding children's mental age. We will not do so. Parent report has shown reliability in past studies, as described in the original proposal, and mental age is only of a concern here to ensure that children are*

> *capable of taking part in an interview; it will not be examined as a "variable" in a quantitative sense.'*

- Similarly, the reviewer asked about 'checking medical records' to check the validity of information from parents. For the same reasons as those in the previous example, we wrote:

> *'The reviewer asks if we will consult medical records regarding diagnosis. We will not do so. Previous research we describe suggests that the issues people with ID have with inpatient care revolve around others' perceptions of their abilities/disabilities/limitations and are not diagnosis specific. We expect any salient aspects of the child's condition (e.g. autistic behaviours) will emerge as their inpatient experience is described.'*

- This reviewer also wrote 'Data analyses are not sufficiently explained.' Obviously what we had written was either not clear enough or lacked clarity for this reviewer. We saw this as an opportunity to add more about feminist poststructuralism. We wrote: *We have elaborated on data analyses in the proposal.'* You can read our elaboration in the 'Methodology' section below.

Responses to Reviewer 2:

- Reviewer 2 was concerned that our sample size of 60 participants (20 in each of 3 groups) was too small. For a qualitative poststructural study, 60 participants is a very large sample. Consequently, in response we pointed out that we expected to reach saturation based on other studies. Therefore, in response we wrote:

> *'This reviewer asks that we justify our sample size of 20 participants for each group to be interviewed (parents, children, nurses). We expect to reach saturation, within each group of participants (parent/child/nurse) through up to 20 interviews, based on previous studies of similar questions based on interviews with parents, nurses and children or adults with intellectual disabilities that have reported sample sizes of 8 to 25.'*

Responses to Reviewer 3:

- Similar to reviewer 1, reviewer 3 also had concerns about 'including information about past history of hospitalizations.' We took this as another opportunity to highlight how a poststructual approach enabled experiences of the past and present to emerge through personal stories. We wrote:

> *'We expect quality of past admissions may be more influential in perceptions than quantity. However, cross-case analyses will examine*

> *the valence and reported number of past hospitalizations at this hospital and others (some children may be admitted to community care hospitals closer to their home in addition to stays at the [health centre] where the study will take place) and how these relate to the personal experience of children and parents.'*

We also provided a common response to all three reviewers regarding feminist poststructural methodology. The following points highlight what we clarified and expanded upon in our methodology section. Specifically, we wrote more about poststructural concepts as well as indicating who on the team was an expert in feminist poststructuralism to ensure rigour and trustworthiness of the study. We wrote that:

- *A more comprehensive explanation of how the concepts of agency and power as relational and situational will guide the data analysis process has been included.*
- *Clarity to research rigour and confidentiality has been added.*
- *More details pertaining to the role of the applicants in data collection and analysis. [Principle investigator] is an expert in the use of feminist poststructuralism and has previously and successfully trained and guided two research assistants in the application of this methodology that were used during interviews and data analysis. The research coordinator will not be left to complete these tasks on [their] own. The applicants will be available for consultation before, during, and after the interviews as well as closely involved in analysing the transcripts as this will be a group effort, (not to be done in isolation).*

Our Elaboration of Methodology and Methods

The examples above clearly indicate that all reviewers were asking for further elaboration on aspects of methodology and methods to be used. In the next section we share our response to satisfy these requests.

A Feminist Poststructural Perspective

To understand the personal, social, and institutional experiences of children with ID, their parents, and nurses who care for children with ID and their families, feminist poststructuralism will be used. Although feminist poststructuralism is methodologically rooted in concepts from women's studies, it moves beyond a narrow focus of gender to include diverse perspectives of abilities, race, ethnicity, class, socio-economic status, culture etc. It supports the belief that people are connected socially and allows for an in-depth examination of personal experiences, relationships, and contextual meanings of relations of power between individuals, society, and institutions. Feminist poststructuralism will offer a unique methodology to understand how interactions between children with ID, parents, and nurses are contextualized personally, institutionally, socially, and politically through a mediation of different discourses in the context of relations of power and empowerment. The concepts of discourse analysis, subjectivity, and agency will unveil a critical understanding of the relationships and practices of children with ID, their parents and nurses.

Discourse analysis incorporates the concepts of language, practices, beliefs and values. For example, there may be different 'discourses' or ways of understanding one's experience of care in the hospital. Individuals viewing the same thing may actually "see" something different depending on their perspective, their culture, or their upbringing. Our understanding of the world is a social construction. Discourses may complement, conflict, or challenge one another. For example, the concepts, practices and experiences of disability or pain may be understood through a medical lens in one discourse. Another discourse may conceptualize pain or disability through personal or intuitive experiences. Discourses may be separate, oppositional, complimentary, or even overlap at times. How a discourse affects or influences a person's beliefs and practices can be understood through their 'lived experiences' and how they tell their stories in a complex and relational manner.

The concept of relations of power is also used in discourse analysis. Power as relational and situational is a shift away from the linear and victim blaming notions of 'power over' and 'powerlessness.' Instead, power is understood to come from within people and is framed by their position and their practices with others[44]. Individuals are not inherently formed static beings having things done to them and neither are institutions inherently bad or controlling. Rather, individuals do have choice, are interactive, influenced by social constructs, complex, and change identities depending on their situations and with whom they interact.

The concepts of subjectivity and agency will also guide the research process. Subjectivity recognizes the importance of understanding who a person is as defined from their own personal location in a certain context. Their subjectivity may change from place to place. For example, an individual may feel like a child, adult, mother, teacher, or nurse, depending on where [they] and who [they are] with. Therefore, it is important to understand a person through their own personal presentation of complex thoughts, beliefs and values. The concept of agency shifts the focus of power to participants. Having agency assumes that individuals have potential control over their lives and the ability to make changes.

The use of feminist poststructuralism values the participants as experts in their own lives, and their description about their experience as truthful and credible sources of data. The use of the concept agency also positions study participants as self-reflexive, conscious of their own social locations, and having the potential to question, challenge and possibly change their own circumstances, while recognizing the oppressive nature of social structures, stereotypes, and ideologies. Each perspective whether it be from the child, parent or nurse is considered to be reliable and trustworthy as it is told from each person's individual perspective and experience. Feminist poststructuralism supports the belief that people are connected socially to their surroundings therefore their experiences will be informed by social stereotypes, societal beliefs and institutional norms. Hence, one can see the importance of paying close attention to language, beliefs, values and practices when analysing each individual's understanding of their experience. Data collected through individual interviews will therefore incorporate multiple layers of understanding that include personal, social, institutional and political aspects. This type of exploration permits greater reflexivity with the data and the participants as well as examination of interactive relations of power between participants situated within social structures of meaning. The 'lived experience' of participants will uncover experiences and individual interpretations told by children with ID, their parents and health care professionals that highlight the tensions and supports of their hospital experience.

Methods

It was also important to clearly articulate how we would proceed with data collection and analysis using a feminist poststructuralist lens. For example, we added more detail and stated that the interviewer would use techniques from a feminist poststructuralist methodology to ensure that the semi-structured interviews were implemented in a non-hierarchical, reciprocal, respectful manner that encourages participants to tell their stories in their own voice, from their own perspective.

Participants were also assured that there were no right or wrong answers and the dialogue was meant to explore different individual meanings of their experience in the hospital. Involvement by the researcher through back-and-forth dialogue as well as accepting what each participant said was truthful is based on the epistemological and ontological stances of feminist and poststructural thought. We believed it helped to foster a trusting relationship and elicit clear and meaningful data. This positive and supportive style of interviewing helped to ensure that children with ID, who were considered to be a vulnerable group, were treated with respect and sensitivity.

Data analysis

he reviewers also asked us to provide more detail and clearly outline how we would specifically analyse the data including who would do it and how long it would take to analyse each interview transcript. We wrote:

> *For example, a one-hour transcript takes an average of 6 hours to ini-tially analyze using discourse analysis. This is then followed by group discussion, further discourse analysis and creating themes. Analysis for one transcript can take up to 18 hours... Emerging discourses and themes will be compared within each interview and then between the interviews. This will include organizing the data using concepts such as discourse, language, subject position, agency and relations of power. Data analysis and collection will occur simultaneously.*

It is crucial to provide enough detail to ensure that reviewers can clearly understand what you are proposing to do, particularly if they do not have experience or knowledge about postmodern/poststructural research. It is also important to indicate that the team has the expertise to do the proposed research.

Establishing the Credentials of Both the Research and the Researcher

An important skill pertaining to the craft of developing a research pro-posal is that of "selling" the research and the researcher(s). The infor-mation requested in the first part of a proposal is used to contextualize both the research and the researcher(s). This information indicates who will work on the project and what role and what responsibilities each person will have. This information will be used to establish the track record or credentials and expertise of the researcher(s) including other grant monies won and publications. It is important to provide clear and complete information that is easy to follow.

Establishing credentials in research draws on often taken-for-granted understandings of either what does or does not "count" in terms of track record or how much each aspect does count in relative terms. The written word (and particularly certain types of written word) is afforded primacy over the spoken word. There is a binary opposition operating here.

Concluding Remarks

In this chapter, we have continued our discussion of proposing postmod-ern/poststructural research by focussing on navigating challenges when writing such research for audiences such as funders, review committees, and individual reviewers of that research especially when those reviewers are not familiar with postmodern/poststructural approaches. Our hope is that you will be able to use the lessons we have learned about navigating these challenges, to assist your own proposal writing and/or your understandings of postmodern/poststructural research.

Postmodern/poststructural research can, and should be, used in practice to make a difference that can deconstruct, question, challenge, and offer new possibilities for change. As researchers we wish you great success and enjoyment as you develop your own research studies.

In the next chapter, the final one of the book, we round off our book length discussion by reflecting on the journey that our discussion in this book has taken us on when exploring postmodern and poststructural approaches. When doing so we offer encouragement for readers to continue their own journey and self-reflection about postmodern and poststructural research.

Notes

1 See Chapter 5 where we discuss the discourses that make up the shape that case notes take and Chapter 4 the discussion of peer reviewed articles as shaped by particular discourses about science.
2 Funding rates for CIHR can be found at the website https://cihr-irsc. gc.ca/e/53104.html

References

Aston, M., Breau, L., & MacLeod, E. (2014a). Understanding the importance of relationships from the perspective of children with intellectual disabilities, their parents, and nurses. *Journal of Intellectual Disability*, *18*(3), 221–237. https://doi.org/10.1177/1744629514538877

Aston, M., Breau, L., & MacLeod, E. (2014b). Diagnoses, labels and stereotypes: Supporting children with intellectual disabilities in the hospital. *Journal of Intellectual Disabilities*, *18*(4), 291–304. https://doi.org/10.1177/1744629514552151

Breau, L., Aston, M., & MacLeod, E. (2016). Education creates comfort and challenges stigma towards children with intellectual disabilities. *Journal of Intellectual Disabilities*, *21*(1), 18–32. https://doi.org/10.1177/1744629516667892

Cheek, J. (2022). The impact of funding on ways qualitative research is thought about and designed. In U. Flick (Ed.), *The SAGE Handbook of Qualitative Research Design*, London, SAGE Publications Ltd., pp. 636–651.

Cheek, J. (2024). Academic survival: Qualitative researchers in the neo-liberal academy. In N.K. Denzin, Y.S. Lincoln, M.D. Giardina, & G.S. Cannella (Eds.), *The SAGE Handbook of Qualitative Research*, 6th Edition.

Koro-Ljungberg M., Douglas E.P., Carlson D., & Therriault D.J. (2015). An unfinished dialogue about problematizing knowledge production in the peer review process. In N.K. Denzin & M.D. Giardina (Eds.), *Qualitative inquiry and the politics of research*. Walnut Creek, CA, Left Coast Press, pp. 27–50.

8 The Journey Forward

Postmodern, Poststructural, and Practice Intersections

In this chapter we will

- Reflect on where the discussion in the book has taken us
- Encourage the reader to continue to explore postmodern and poststructural approaches and ways to implement them in their own research into practice
- Present reflections from scholars both experienced and novice about their use of postmodern and poststructural approaches in their research
- Encourage the reader to be self-reflexive both about their own research and how they use postmodern and poststructural approaches to inform their practice

---

Introduction: The Final Chapter But Not the Final Word

Reading this book can be likened to undertaking an exploratory journey. This journey has focused on two key interrelated areas: what postmodern and poststructural approaches are, and whether these approaches can be used in research designed to inform practice areas such as those in health care or educational settings. When exploring these two interrelated areas, the way in which practically oriented research drawing on these approaches can be developed, and carried out, has been an integral part of the discussion.

No easy answers have presented themselves to the questions and issues that have arisen in this exploratory journey. As we have seen, the very definitions and understandings of postmodern and poststructural approaches are contested, and so is their relevance to practice. However, we have been able to map out some guiding principles about, and parameters for, understanding what postmodern and poststructural approaches are, and how they might be used to inform practice. By so doing we hope to provide

DOI: 10.4324/9780429053764-8

readers with a way of avoiding becoming bogged down in the mire of theoretical and methodological ambiguity, vagueness, and assumed understandings about the nature, conduct, and purpose of a research endeavour informed by postmodern and poststructural thought.

> If you would like to get a sense of where this journey has taken us reading the abstracts of each chapter progressively one after the other provides a good overview of the theoretical, methodological, and substantive elements of this journey of exploration we have undertaken.

The Journey Continues: Where to From Here?

So where to now in terms of our exploratory journey? In the last chapter of a book it is customary to reach some sort of "final" statement or conclusion. Such a conclusion usually pulls together the threads of the discussion that have been systematically developed in the preceding pages of that book. Having read such a conclusion, the reader is left with a sense of closure, a sense that a journey has been completed, and a certain destination reached.

The conclusion to this book is not like that. In many ways it would be antithetical to the discussion of postmodern and poststructural approaches to attempt to arrive at some sort of defined end point. Instead, in keeping with the philosophy of these approaches, this final chapter aims to both round off the discussion of where we have been, but at the same time open up ideas and further avenues for exploration; for example, ways in which postmodern and poststructural approaches have been, and might be, used in conjunction with other theories to explore contemporary practice issues. Thus, in many ways this chapter marks the start of what we hope will be an ongoing journey of discovery for you, the reader.

As you continue your journey of exploration about postmodern and poststructural approaches to research you will encounter many examples of how researchers have combined postmodern and poststructural approaches with other theories in the one study. This opens new and different possibilities for research that can inform practice. The next sections describe different ways that scholars have done this.

Megan's Journey Using Poststructural Approaches in Research and Practice

I was first introduced to postmodern and poststructural approaches as a Master of Education student. Foucault (1982) was one of the primary poststructural scholars informing my work along with feminist scholars

(Butler, 1992; Scott, 1992; Weedon, 1987), Black scholars (bell hooks, 1984; Collins, 1990), and other critical social scholars (Spivak,1988; Freire, 1986) including my supervisors Roger Simon (1991) and Magda Lewis (1986). After initially becoming familiar with a poststructural approach I was then excited to see the relevance of using a feminist perspective in combination with poststructural thought. Weedon wrote about how Foucauldian ideas and feminist approaches could be brought together.

> Both identify the body as the site of power, that is, as the locus of domination through which docility is accomplished and subjectivity constituted.... Despite their seemingly different objectives, then, feminist and Foucauldian analyses come together in the ways they have attempted to dismantle existing but heretofore unrecognized modes of domination.
>
> (Weedon, 1987, p. x)

While some scholars criticized Foucault for not being political enough (Hartsock, 1989), I argued that postmodern and poststructural approaches *were* political with a focus on power and deconstruction, and therefore used poststructural thought in my research to support a social justice perspective in health practice. For example, the poststructural concept of binary relations has been invaluable in my research. When not questioned or challenged, binary opposites can perpetuate reductive subject positions, such as doctor/patient and man/woman. In chapters 3 and 4 we wrote about the importance of identifying binary opposites and using this as an opportunity to deconstruct moments of tension and struggle through relations of power.

For me personally, throughout my Masters, PhD, and ongoing academic research, postmodern/poststructural and feminist approaches have offered a unique way to understand how social and institutional discourses are constructed and negotiated through relations of power. I have used postmodern and poststructural approaches throughout all of my research to understand how power was used by and empowered people.

> Not only do participant experiences show us the challenges of negotiating relations of power, but they also provide solutions because they have lived it. Questioning and disrupting everyday practices has the potential to ensure we are attending to different discourses in an equitable way.
>
> (Aston, 2016, p. 2265)

Postmodern and poststructural approaches have been invaluable to all of my scholarly teaching and research. In the box below you will find a listing of my research articles that span a variety of topic areas. All of the studies used postmodern and poststructural approaches with a focus on social justice. I have thoroughly enjoyed leading and participating in all of the different research studies over the years and believe we were able to make a difference in each practice area. I also share one publication that outlines my personal journey teaching postmodern and poststructural approaches and links to two videos we developed based on research findings.

Research studies conducted by Megan using a poststructural approach include:

- Postpartum care, mothers, and public health nurses (Aston, 2002, 2008; Aston et al., 2006, 2014, 2015, 2016, 2018, 2019, 2020, 2021; Joy et al., 2020; Benoit et al., 2023; MacLeod et al., 2023)
- Maternity care in Tanzania (Mselle et al., 2016; Kohi et al., 2017; Mbekenga et al., 2018)
- Queer birthing, children with intellectual disabilities (Aston et al., 2014a, 2014b, 2019; Breau et al., 2016; Vanderlee et al., 2020)
- Bereavement care, (MacConnell et al., 2012)
- Obesity management (Aston et al., 2011; Kirk et al., 2014)
- LGBTQ+ compassion (Joy et al., 2022).

You can read more about how Megan teaches postmodern and poststructural approaches in the following article:

Aston, M. (2016). Teaching feminist poststructuralism: Founding scholars are still relevant today. *Creative Education*, 7(15), 2251–2267. doi:10.4236/ce.2016.715220

Megan and her team created a video called *Mindful Matters* to present research findings that used postmodern, poststructural, and feminist approaches to understand the hospital experiences of children with intellectual disabilities (ID), their parents and nurses who cared for them. The video has been used for teaching purposes in workshops with students. As you watch the video, try to identify the ways in which they purposefully chose to share the findings using postmodern, poststructural, and feminist approaches. https://vimeo.com/163234949

Megan and others also created another video called *Mommy Dialogues*. Have a look and see how they incorporated FPS https://www.youtube.com/watch?v=G0Z69omZSAY

Gaining Inspiration from Students' Journeys

In the discussion to follow we highlight examples of how students who at the time were relatively new to postmodern and poststructural inspired research, used these approaches in their studies, sometimes in combination with other theories. We hope that these practical and frank examples will inspire and encourage readers – especially those with less experience in using these approaches.

Example 1: **Emma Vanderlee RN MScN (former student) – reflecting on her first experience of using Feminist Poststructuralism (written 1 year after being introduced to FPS as a Masters student). (Emma is presently a PhD student)**

I started my journey of learning about feminist poststructuralism (FPS) a year ago, after I was accepted in the Masters of Science in Nursing program at Dalhousie University. My thesis supervisor, an expert in FPS helped me find a research job over the summer where I had the opportunity to use this methodology to complete one of Megan's research studies. The first step with her assistance was for me to gain an understanding of what FPS was. I started by reading several books. The authors included Weedon, Foucault, and Butler. I clearly remember reading pages and words, over and over again, and questioning if I truly understood the English language. It was a steep learning curve. I had not anticipated coming into my Master's program, and being met with such an overwhelming experience. It was humbling as I did not easily understand the meaning behind the thoughts and concepts of authors such as Foucault. I understood my thesis advisor's rationale for exploring and comprehending the fundamental theories and intricacies of FPS. However, delving into the underlying connections between feminism, social constructs, and this methodology required meticulous examination and a substantial time investment.

My initial learning stimulated a curiosity to know more about how FPS inspired research approaches. I also saw how unconsciously these concepts governed multiple aspects of my life, and this really helped push me to want to continue learning about FPS. But I was still struggling to fully and practically understand how to apply FPS in a research setting and to be able to talk about this and explain it clearly.

I believe that my initial struggle as a student trying to learn about FPS largely arose because I was reading these complex texts looking for a recipe, technique, or method for how I could use or apply FPS in my summer research internship. In hindsight, I should have been reading these original texts more contextually and historically to understand why FPS was formed and how it is developing over time. A primary reason for the need for multiple readings of the texts was not only the difficulty of the theories but also, if not more so, a consequence of the lens I was using to read – namely searching for practical recipes and techniques. For the previous four years as an undergraduate nursing student, I had learned to read practically rather than philosophically, historically, holistically, and contextually. This means, that while it seems prudent for academic teachers to assign original texts to students to help them learn it is important students also understand how to read such texts and know why this reading is important to the student trying to learn about FPS.

By the time I read and re-read the four assigned books (which took all summer), I had begun to recognize that I needed to shift my lens for reading, I had started my course work for my Masters' program. Using a different lens for reading, one not focused on searching for practical recipes and techniques, I found I was better able to understand the complexities, rationale, history, and contexts for using FPS as a methodology. I began to shift my viewpoint in class and at conferences to ways that different methodologies can be applied to research. I also began to better understand how a vast body of knowledge such as FPS can help answer a research question. I began to grasp the potential of versatile applications of FPS, allowing my imagination to formulate innovative ways to integrate it into various research scenarios.

As a novice student, with a better understanding of how to read texts about complex concepts, I also found it easier to learn about other research methodologies. In turn, helping me to better understand the foundational concepts of FPS such as discourses, subjectivity and agency, language, binary opposites, power, etc. It shifted how I read other texts over the summer and during my first year of study. I am now completing my first year of the program and I have the confidence to speak to others about the methodology. My goal for this summer is to reread the books, through a new lens, to further grasp, re-new and re-situate my understanding of FPS.

Example 2: Dr. Phillip Joy reflecting on his experience of using Queer and Poststructural Approaches in his doctoral work (written while still a doctoral student – Phillip is now an Assistant Professor)

Poststructuralism is used in my research as a critical theoretical philosophy that underpins and supports my use of arts-based methods, such as photovoice. Like poststructuralism art can be used to critique fundamental assumptions of knowledge, power, and ways of knowing oneself. Both can disrupt fundamental assumptions within health research that truth and knowledge are independent of human experiences, emotions, and creativity.

My research often explores the experiences of people within LGBTQ communities. This opens up the opportunity to interweave poststructuralism with queer theory: a theory derived from gay and lesbian identity politics. The interweaving of these two philosophical perspectives gives my work a political and advocacy aspect that can be expressed through art and art shows. Queer theory allows for the reconceptualization of binary notions of identity, gender, and sexuality which is often useful when exploring the tenets of poststructuralism: language, knowledge, and power. Poststructuralism used with queer theory and reflected through art gives people the possibility of questioning, exploring, and, ultimately, changing existing social and cultural arrangements of gender and the opportunity to resist and subvert these arrangements.

My current doctoral work explores how social discourses and knowledge shape the beliefs, values, and practices about food and bodies, and the influence this has on the health of gay men. I use images, to challenge and disrupt the dominant notions about bodies and try to contribute to social changes through the expression of new, hidden, or silenced discourses and knowledge about the bodies of gay men. This was further explored in an article I published entitled *Constituting the Ideal Body: A Poststructural Analysis of "Obesity" Discourses among Gay Men*. My research, by the integral use of poststructuralism, queer theory, and art, aims to shift understanding of gay men's bodies and to acknowledge the complexities within their nutritional practices and experiences.

Joy, P. & Numer M. (2018) Constituting the Ideal Body: A poststructural analysis of 'obesity' discourses among gay men. *Journal of Critical Dietetics*. 4(1) ISSN 1923-1237

Queer poststructural approach used to develop a comic book Rainbow Reflections: Body Image Comics for Queer Men. By Phillip Joy

Building on his PhD study Phillip developed a comic series with researchers and artists around the world focusing on gay men's body image.   https://www.dal.ca/news/2019/07/12/framing-the-issue--phd-student-tackles-queer-men-s-body-image-in.html?utm_source=Today@Dal&utm_medium=email&utm_campaign=dalnews

---

Example 3: Dr. Rachel Ollivier RN PhD When Rachel was a PhD student and before she completed her research she wrote the following (Rachel recently attained her PhD in 2022)

Feminist poststructuralism is highly relevant and applicable to the topic of postpartum sexual health in that it provides concepts that lend themselves to the critique of discourse and meaning. As an example, my research will aim to explore the meaning of postpartum sexual health for postpartum individuals living in Nova Scotia, Canada.

Poststructuralism has informed much of my work thus far as I begin to identify and critique how language, gender, agency, subjectivity, and relations of power influence how postpartum sexual health is experienced, viewed, or interpreted by postpartum women or wider social discourse. In using feminist post-structural discourse analysis for my research, I will first identify the beliefs, values, and practices present in women's experiences to subsequently identify meaning within those experiences. I am intentional in using feminist poststructuralism because it allows me to explore questions that arise both in research and in clinical practice. In other words, it goes deeply enough to address institutional, social, or political relations of power while providing direction for solutions, implications, or other possibilities in health and health care.

I admire feminist poststructuralism because it's political and doesn't shy away from controversial or misunderstood topics, but rather provides a means to name and deconstruct issues. For me, it's important to answer "So what?"– What do we do about it? Feminist poststructuralism has remained relevant for me because I think of it as a toolbox. More specifically, I have found Foucault and Weedon to be especially helpful to my work in their concepts of agency, discourse, and subjectivity, with Foucault addressing these concepts specifically in relation to sexuality. If you want to read more about Rachel's work please see:

Ollivier, R., Aston, M., & Price, S. (2018). Let's talk about sex: A feminist post-structural approach to addressing sexual health in the health care setting. *Journal of Clinical Nursing, 28*(5). doi: 10.1111/jocn.14685

When researching material for this book Megan also talked to some of her research colleagues about their experiences of, and perceptions about using postmodern and poststructural perspectives in their research. The result was many insightful reflections on different aspects of the challenges, but also rewards, of engaging with these approaches in their research. While we could not include them in this chapter length discussion (despite many attempts), we agreed that they were too valuable not to share with readers, so we decided to include parts of those contributions in a special section called Coda. Similar to a Coda in a musical work, our Coda provides a finale that connects to and rounds out the ideas that went before. These reflections provide frank insights into the reality of navigating issues and challenges that you may encounter when putting postmodern and poststructural approaches into practice.

The reflections and examples are from:

Dr. Danielle Macdonald
Dr. Sara Kirk
Dr. Judy MacDonald
Dr. Matt Numer
Dr. Lisa Goldberg
Evelyn Abudulai

You can find them on pages 123–133

The Journey Forward

Giroux (1992) has promoted what he terms "border crossing" (p. 22) in order to create "borderlands" (p. 22) or "alternate public spaces" (p. 22) where it is possible to rewrite "histories, identities and learning possibilities" (p. 30). This chapter, like the rest of the book, has promoted the crossing of borders when on exploratory journeys of postmodern and poststructural approaches in order to create, new spaces from which to view and research aspects of practice. Pushing beyond the constraints of the discursively constructed and maintained borders of taken-for-granted understandings of what practice *is*, opens up possibilities for what that practice *might be*. Yet more borders are crossed, and new spaces opened up, when postmodern/poststructural thinking and research about practice are combined with different theories across a variety of disciplines.

Postmodern and poststructural approaches have a place in all areas of research, including those related to practice – as do a myriad of other theoretical perspectives and methodological approaches. "The knowledge claims of each are potentially adequate and appropriate in different circumstances and given different purposes" (Rosenau, 1994, p. 96). However, at all times, the rationale for choosing a particular research approach (including postmodern and poststructural ones) rather than other possibilities must be stated clearly. The strengths and limitations of the various approaches must be recognized, as must the role that theoretical and methodological frames play in shaping research understandings, research undertakings, and conclusions reached from that research.

Postmodern and poststructural approaches, like any other approach to research, represent certain views of reality. Put another way, postmodern and poststructural approaches themselves are discursive constructions, drawing on certain knowledge claims to give them presence. Thus, they are open to the same sort of challenge and scrutiny as any other theoretical frame. Their use must be considered and able to be justified – both with respect to why and how they were used. Therefore, throughout the book we have emphasized that in order for the use of postmodern and poststructural approaches to be considered, justifiable and justified, development of reflexivity on the part of the researcher using those approaches is required.

Berger (2015) writes about the importance of researcher reflexivity in order to expose how a researcher's positioning of their self and the beliefs they hold that arise from that self, can affect the way they engage with participants or texts in research:

> (R)esearchers need to increasingly focus on self-knowledge and sensitivity; better understand the role of the self in the creation of knowledge; carefully self monitor the impact of their biases, beliefs, and personal experiences on their research… It means turning of the researcher lens back onto oneself to recognize and take responsibility for one's own situatedness within the research and the effect that it may have on the setting and people being studied, questions being asked, data being collected and its interpretation. As such, the idea of reflexivity challenges the view of knowledge production as independent of the researcher producing it and of knowledge as objective.
>
> (p. 220)

Likewise Lumsden (2019) writes about the importance of using reflexivity in our postmodern/poststructural inspired research.

> A reflexive approach enables us to be conscious of the social, ethical, and political impact of our research; the central, fluid, and changing nature/s of power relations (with participants, gatekeepers, research funders, etc.); and our relationships with the researched, aspects which diffractive methodologies overlook.
>
> (p. 6)

Such reflexive situating of oneself as researcher includes acknowledging and reflecting on the effect on the research of relevant personal characteristics of the researcher and/or the context the researcher comes from. This can include the researcher's gender, race, affiliations, age, sexual orientation, immigration status, personal experiences, linguistic tradition, beliefs, biases, preferences, theoretical, political, and ideological stances, and emotional responses to participants (Bradbury-Jones, 2007; Finlay, 2002; Hamzeh & Oliver, 2010; Horsburgh, 2003; Kosygina, 2005; Padgett, 2008; Primeau, 2003).

At times such self-reflexivity can be confronting work. As in any journey of exploration and discovery, exploring and using, postmodern and poststructural approaches poses challenges, many of which we have touched on in this book. Nevertheless, the potential offered by such a journey for opening up and rewriting histories, identities, and research possibilities in practice, (to paraphrase Giroux, 1992), is enormous, unsettling, and well worth the risk to challenge aspects of reality we have come to take for granted.

This book has been a small part of the journey forward. We hope it encourages readers to continue this journey of exploration. As you do so always remember – postmodern and poststructural approaches and thinking *can* be used in research about practice; and *can* influence and improve that practice.

References

Agger, B. (1992). *Cultural Studies as Critical Theory*, London, Falmer Press.

Armstrong, D. (1985). The subject and the social in medicine: An appreciation of Michel Foucault. *Sociology of Health and Illness, 7*(1), 108–117.

Aston, M. (2002). Learning to be a normal mother. Empowerment and pedagogy in postpartum classes. *Public Health Nursing, 19*(4), 284–293. https://doi.org/10.1046/j.1525-1446.2002.19408.x

Aston, M. (2008). Public health nurses as social mediators: Using feminist poststructuralism to guide practice with new mothers. *Nursing Inquiry, 15*(4) 280–288. https://doi.org/10.1111/j.1440-1800.2008.00408.x

Aston, M. (2016). Teaching feminist poststructuralism: Founding scholars are still relevant today. *Creative Education, 7*(15), 2251–2267. https://doi.org/10.4236/ce.2016.715220

Aston, M., Breau, L., & MacLeod, E. (2014a). Diagnoses, labels and stereotypes: Supporting children with intellectual disabilities in the hospital. *Journal of Intellectual Disabilities, 18*(4), 291–304. https://doi.org/10.1177/1744629514552151

Aston, M., Breau, L., & MacLeod, E. (2014b). Understanding the importance of relationships from the perspective of children with intellectual disabilities, their parents and nurses. *Journal of Intellectual Disability, 18*(3), 221–237. https://doi.org/10.1177/1744629514538877

Aston, M., Etowa, J., Price, S., Vukic, A., Hart, C., MacLeod, E., & Randel, P. (2016). Public health nurses and mothers challenge and shift the meaning of health outcomes. *Global Qualitative Nursing Research, 3*, 1–10. https://doi.org/10.1177/2333393616632126

Aston, M., Meagher-Stewart, D., Sheppard-LeMoine, D., Vukic, A., & Chircop, A. (2006). Family health nursing and empowering relations. *Pediatric Nursing, 32*(1), 61–67. Retrieved from http://ezproxy.library.dal.ca/login?url=https://search-proquest-com.ezproxy.library.dal.ca/docview/199457220?accountid=10406

Aston, M., Price, S., Etowa, J., Vukic, A., Young, L., Hart, C., MacLeod, E., & Randel, P. (2014). Universal and targeted early home visiting: Perspectives of Public Health Nurses and Mothers. *Nursing Reports, 4*(1). https://doi.org/10.4081/nursrep.2014.3290

Aston, M., Price, S., Etowa, J., Vukic, A., Young, L., Hart, C., MacLeod, E., & Randel, P. (2015). The power of relationships: Exploring how public health nurses support mothers and families during postpartum home visits. *Journal of Family Nursing, 21*(1), 11–34. https://doi.org/10.1177/1074840714561524

Aston, M., Price, S., Hunter, A., Sim, M., Etowa, J., Monaghan, J., Paynter, M. (2020). Second opinions: Negotiating agency in Online Mothering Forums. *Canadian Journal of Nursing Research.* https://doi.org/10.1177/0844562120940554

Aston, M., Price, S., Kirk, S., & Penney, T. (2011). More than meets the eye. Feminist poststructuralism as a lens towards understanding obesity. *Journal of Advanced Nursing, 68*(5), 1187–1194. https://doi.org/10.1111/j.1365-2648.2011. 05866.x

Aston, M., Price, S., Monaghan, J., Sim, M., Hunter, A., & Little, V. (2018). Navigating and negotiating information and support: Experiences of first time mothers. *Journal of Clinical Nursing* [shared first author] https://doi.org/10.1111/jocn.13970

Aston, M., Price, S., Paynter, M., Sim, M., Monaghan, J., Jefferies, K., & Ollivier, R. (2021). Mothers' experiences with child protection services: Using qualitative feminist poststructuralism. *Nursing Reports, 11*(14) 913–928. https://doi.org/10.3390/nursrep11040084

Aston, M., Sweet, K., Price, S., McAfee, E., Filliter, J., Sheriko, J., Monaghan, J., Vanderlee, E., McGrath, P., Bye, A., & Walls, C. (2019). Snap Shot: Achieving better care through a one page Personal Health Profile. *Journal of Intellectual Disabilities*. https://doi.org/10.1177/1744629519873503

Ball, S. (1990). Introducing Monsieur Foucault. In S.J. Ball (Ed.), *Foucault and Education: Disciplines and Knowledge*, London, Routledge, pp. 1–8.

Bartky, S. (1988). Foucault, Femininity, and the Modernization of Patriarchal Power. In I. Diamond & L. Quinby (Eds.), *Feminism and Foucault: Reflections on Resistance*, Boston, Northeastern University Press, pp. 61–68.

Benoit, B., Aston, M., Price, S., Iduye, D., Sim, M., Ollivier, R., Joy, P., & Akbari, N.A. (2023). Mothers' access to social and health care systems support during their infants' first year during the COVID-19 pandemic: A qualitative feminist poststructural study. *Nursing Reports*. https://doi.org/10.3390/nursrep13010038

Berg, T. (2020). Manifestations of surveillance in private sector dance education: The implicit challenges of integrating technology. *Research in Dance Education*, 135–152. https://doi.org/10.1080/14647893.2020.1798393

Berger, R. (2015). Now I see it, now I don't: Researcher's position and reflexivity in qualitative research. *Qualitative Research, 15*(2), 219–234.

Bradbury-Jones, C. (2007). Enhancing rigour in qualitative health research: Exploring subjectivity through Peshkin's I's. *Journal of Advanced Nursing, 59*(3), 290–298.

Breau, L., Aston, M., & MacLeod, E. (2016). Education creates comfort and challenges stigma towards children with intellectual disabilities. *Journal of Intellectual Disabilities, 21*(1), 18–32. https://doi.org/10.1177/1744629516667892

Brown, C., & Seddon, J. (1996). Nurses, doctors and the body of the patients: Medical dominance revisited. *Nursing Inquiry, 3*(1), 30–35.

Burrell, G. (1988). Modernism, post modernism and organizational analysis 2: The contribution of Michel Foucault, *Organizational Studies, 9*, 221–235.

Butler, J. (1992). Contingent foundations: Feminism and the question of "Postmodernism." In J. Butler & J. Scott (Eds.), *Feminists Theorize the Political* New York and London: Routledge, Chapman and Hall, Inc, p. 3.

Cheek, J., & Rudge, T. (1994). Inquiry into nursing as textually mediated discourse. In P. Chinn (Ed.), *Advances in Methods of Inquiry for Nursing*. Gaithersburg, Aspen Publishers, pp. 59–67.

Collins, R. (1990). Cumulation and anti-cumulation in sociology. *American Sociological Review, 55*, 462–463.

Davies, B. (1989). *Frogs and Snails and Feminist Tales*. Sydney, Allen and Unwin.

Dean, M. (1994). *Critical and Effective Histories: Foucault's Methods and Historical Sociology*, London, Routledge.

Einboden, R. (2020). SuperNurse? Troubling the hero discourse in COVID times. *Health, 24*(4), 343–347. https://doi.org/10.1177/1363459320934280

Finlay, L. (2002). Negotiating the swamp: The opportunity and challenge of reflexivity in research practice. *Qualitative Research, 2*(2), 209–230.

Foster, H. (1985). Postmodernism: A preface. In *Postmodern Culture*. London, Pluto Press, pp. ix–xvi.

Foucault, M. (1975). *The Birth of the Clinic*. New York, Vintage Books.

Foucault, M. (1977). *Discipline and Punish*. London, Tavistock.

Foucault, M. (1979). Governmentality, *I & C, 5*, 5–21.

Foucault, M. (1980). *Power/Knowledge*, C. Gordon (Ed.), Brighton, Harvester Press.

Foucault, M. (1982). Afterword: The subject and power. In H. Dreyfus & P. Rabinow (Eds.), *Michel Foucault: Beyond Structuralism and Hermeneutics*, Chicago, University of Chicago Press, pp. 208–226.

Foucault, M. (1984). The order of discourse. In M. Shapiro (Ed.), *Language and Politics*, Oxford, Basil Blackwell, pp. 108–138.

Fox, N. (1993). *Postmodernism, Sociology and Health*. Toronto, University of Toronto Press.

Fox, N. (1994). Anaesthetists, the discourse on patient fitness and the organisation of surgery. *Sociology of Health and Illness, 16*(1), 1–18.

Freire, P. (1986). *Pedagogy of the Oppressed*. New York, Continuum.

Gilbert, T. (1995). Nursing: Empowerment and the problem of power. *Journal of Advanced Nursing, 21*, 865–871.

Giroux, H.A. (1992). Literacy, pedagogy, and the politics of difference. *College Literature, 19*(1), 1–11.

Hacking, I. (1991). The archaeology of Foucault. In D. Hoy (Ed.), *Foucault: A Critical Reader*, Oxford, Basil Blackwell, pp. 27–40.

Hamzeh, M.Z., & Oliver, K. (2010). Gaining research access into the lives of Muslim girls: Researchers negotiating muslimness, modesty, inshallah, and haram. *International Journal of Qualitative Studies in Education, 23*(2), 165–180.

Hartsock, N. (1989). Foucault on power: A theory for women? In L. Nicholson (Ed.), *Feminism/Postmodernism*, Routledge, pp. 157–175.

Hooks, B. (1984). *Black Women Shaping Feminist Theory*. ProQuest Information and Learning.

Horsburgh, D. (2003). Evaluation of qualitative research. *Journal of Clinical Nursing, 12*(2), 307–312.

Hoskin, K. (1990). Foucault under examination: The crypto-educationalist unmasked. In S. Ball (Ed.), *Foucault and Education: Disciplines and Knowledge*. London, Routledge, pp. 29–53.

Hoy, D. (1991a). Introduction. In D. Hoy (Ed.), *Foucault: A Critical Reader*. Oxford, Basil Blackwell, pp. 1–26.

Hoy, D. (1991b). Power, repression, progress: Foucault, Lukes and the Frankfurt School. In D. Hoy (Ed.), *Foucault: A Critical Reader*. Oxford, Basil Blackwell, pp. 123–148.

Johnson, J.L. (1994). A dialectical examination of nursing art. *Advances in Nursing Science, 17*, 1–14.

Joy, P., Aston, M., Price, S., Sim, M., Ollivier, R., Benoit, B., Akbari Nassaji, N., & Iduye, D. (2020). Blessings and curses: Exploring the experiences of new mothers during COVID-19 pandemic. *Nursing Reports, 10*(2), 207–219. http://dx.doi.org/10.3390/nursrep10020023

Joy, P., & Numer M. (2018). Constituting the Ideal Body: A poststructural analysis of 'obesity' discourses among gay men. *Journal of Critical Dietetics, 4*(1) ISSN 1923-1237.

Joy, P., Thomas, A., & Aston, M. (2022). Compassionate discourses: A qualitative study exploring how compassion can transform healthcare for 2SLGBTQ+ people. *Qualitative Health Research.* http://dx.doi.org/10.1177/10497323221110701

Kirk, S.F.L., Price, S., Penney, T.L., Rehman, L., Lyons, R., Piccinini-Vallis, H., Vallis, T.M., & Aston, M. (2014). Blame, shame and lack of support. A multilevel study of obesity management. *Qualitative Health Research* [name at end denotes second author] https://doi.org/10.1177/1049732314529667

Kohi, T., Aston, M., Macdonald, D., Mbekenga, C., Mselle, L, Price, S., Murphy, G.T., Murphy, W.M., O'Hearn, S., & Jefferies, K. (2017). Saving lives with caring assessments: How Tanzanian nurse-midwives and obstetricians negotiate postpartum practices. *Journal of Clinical Nursing, 26*(23–24), 5004–5015. https://doi.org/10.1111/jocn.14000 [shared first author]

Kosygina, L.V. (2005). Doing gender in research: Reflection on experience in field. *The Qualitative Report, 10*(1), 87–95.

Kress, G. (1985). *Linguistic Processes in Socio-cultural Practice.* Victoria, Deakin University Press.

Lewis, M., & Simon, R. (1986). A discourse not intended for her: Learning and teaching within patriarchy. *Harvard Educational Review, 56*(4), 457–473.

Long, J. (1992). Foucault's clinic. *Journal of Medical Humanities, 13*(3), 119–138.

Lumsden, K. (2019). *Reflexivity: Theory, Method, and Practice.* Routledge.

MacConnell, G., Aston, M. Randel, P., & Zwaagstra, N. (2012). Nurses' experiences providing Bereavement follow-up: An exploratory study using feminist poststructuralism. *Journal of Clinical Nursing, 22*(7–8), 1094–1102. https://doi.org/10.1111/j.1365-2702.2012.04272.x

MacLeod, A., Aston, M., Price, S., Stone, K., Ollivier, R., Benoit, B., Sim, M., Marcellus, L., Jack, S., Joy, P., Gholampourch, M., & Iduye, D. (2023). "There's an etiquette to Zoom that's not really present in-person": A qualitative study showing how the mute button shapes virtual postpartum support for new parents. *Qualitative Health Research.* Accepted.

Mani, L. (1990). Multiple mediations: Feminist scholarship in the age of multinational reception. *Feminist Review, 35*, 24–41.

May, C., Dowrick, C., & Richardson, M. (1996). The confidential patient: The social construction of therapeutic relationships in General Medical Practice. *Sociological Review, 44*(2), 187–203.

Mbekenga, C., Aston, M., Kohi, T., Macdonald, D., Mselle, L., Price, S., Murphy, G.T., Murphy, W.M., O'Hearn, S., & Jefferies, K. (2018). How Tanzanian nurse midwives, and obstetricians develop postpartum relationships with women. *International Journal of Childbirth, 8*(1), 41–53. https://doi.org/10.1891/2156-5287.8.1.41

Miller, P., & Rose, N. (1990). Governing economic life. *Economy and Society, 19*(1), I–31.

Mselle, L., Aston, M., Kohi, T., Macdonald, D., Mbekenga, C., White, M., Price, S., Murphy, G., Tomblin, M., O'Hearn, S., & Jefferies, K. (2016). The challenges of providing postpartum education in Dar es Salaam, Tanzania: Narratives of nurse-midwives and obstetricians. *Qualitative Health Research, 27*(12), 1792–1803. https://doi.org/10.1177/1049732317717695 [shared first author]

Ollivier, R., Aston, M., & Price, S. (2018). Let's talk about sex: A feminist post-structural approach to addressing sexual health in the health care setting. *Journal of Clinical Nursing, 28*(5). https://doi.org/10.1111/jocn.14685

Padgett, D.K. (2008). *Qualitative Methods in Social Work Research* (2nd ed.). Thousand Oaks, CA.

Price, S., Aston, M., Monaghan, J., Sim, S., Tomblin Murphy, G., Etowa, J., Pickles, M., Hunter, A., Little V. (2017). Maternal knowing and social networks: Understanding first time mothers' search for information and support through online and offline social networks. *Qualitative Health Research, 28*(10), 1552–1563. https://doi.org/10.1177/1049732317748314 [Shared first author]

Price, S., Aston, M., Rehman, L., Lyons, R., & Kirk, S. (2015). Feminist post-structural analysis of obesity management: A relational experience. *Clinical Nursing Studies, 3*(3), 76. https://doi.org/10.5430/cns.v3n3p76

Primeau, L.A. (2003). Reflections on self in qualitative research: Stories of family. *The American Journal of Occupational Therapy, 57*(1), 9–16.

Rosenau, P. (1994). Revitalizing sociology: Post modern perspectives on methodology. *Current Perspectives in Social Theory, 14*, 89–99.

Scott, J. (1992). Experience. In J. Butler & J. Scott (Eds.), *Feminists Theorize the Political*. London, Routledge, Chapman and Hall Inc, pp. 22–40.

Searle, J., Goldberg, L., Aston, M., & Burrow, S. (2017). Accessing new understandings of trauma-informed care with queer birthing women in rural Nova Scotia. *Journal of Clinical Nursing, 26*(21–22), 3576–3587. https://doi.org/10.1111/jocn.13727

Simon, R. (1991). *Teaching against the Grain: Texts for a Pedagogy of Possibility*. Toronto, OISE Press.

Spivak, G.C. (1988). Can the Subaltern Speak? In C. Nelson & L. Grossberg (Eds.), *Marxism and the Interpretation of Culture*. Chicago: University of Illinois Press, pp. 271–313.

Vanderlee, E., Aston, M., Turner, K., McGrath, P., & Lach, L. (2020). Patient oriented research: A qualitative study of research involvement of parents of children with neurodevelopmental disabilities. *Journal of Intellectual Disabilities*. https://doi.org/10.1177/1744629520942015

Weedon, C. (1987). *Feminist Practice and Post Structuralist Theory*. London, Basil Blackwell.

Coda

Colleagues Reflections on the Use of Postmodern and Poststructural Approaches

In this section colleagues shared their frank reflections about their use of postmodern and poststructural approaches in their research sometimes combined with other theories. Some colleagues wrote their responses, and others shared their ideas through informal discussions with Megan.

Dr. Danielle Macdonald PhD RN Had Recently Graduated with Her PhD When Writing This. She Is Now an Assistant Professor at Queen's University Canada

Here are some excerpts from when Danielle wrote about her experiences using poststructural and feminist approaches.

> **Excerpt 1:** I have used feminist poststructuralism as a methodology for research and have primarily relied on the works of Weedon (1987), Foucault (1975, 1982) and Cheek (1999) to inform my understanding of feminist poststructuralism. In reviewing Weedon's construction of feminist poststructuralism, I have come to argue for the need to identify the feminist approach being used with poststructuralism. Although feminism has historically been referred to as a singular theory, scholars and advocates have identified multiple, ever-changing theories where the plural 'feminisms' is a more inclusive term for the variety of feminist approaches and theories that have been and will be constructed. This pluralized understanding of feminisms shares a fluidity of meanings and understandings with poststructuralism.
>
> **Excerpt 2:** In my research, I have situated myself within an intersectional feminist poststructuralist perspective. The intersectional feminist perspective has been informed by the works of hooks

DOI: 10.4324/9780429053764-9

(1984), Collins (1998), Collins and Bilge (2020), Crenshaw (2017), and others. Intersectionality as a theory extends the feminist perspectives primarily concerned with gender to include how the intersections of gender with different subject positions such as class, race, and education, influence our lives socially, politically, individually (Collins & Bilge, 2020). The notion that there are many subject positions that interact with gender is not new, and the core ideas of intersectionality can be clearly seen in the works of many Black feminists, as well as other marginalized groups of women (Collins & Bilge, 2020).

Excerpt 3: With the integration of intersectionality as an identified feminist theory in feminist poststructuralism, the intersection of additional subject positions such as race and class are made visible within the contexts of social, historical, and institutional discourses. It is also important to understand that there is not one way to understand or apply intersectional feminist poststructuralism. Similar to poststructuralism, feminist poststructuralism generally, and intersectional feminist poststructuralism in particular are constructions which are fluid and dynamic depending on the people who engage with these theories and the historical, social, and institutional discourses and contexts that influence the meanings and our understandings of these approaches. Therefore, I am not essentializing intersectional feminist poststructuralism as a singular approach or theory, more specifically I am suggesting that the way that I have integrated the concepts and ideas of both is one way of engaging with feminist poststructuralism, but it's not the only way.

Excerpt 4: The privileging of postpositivist ideas and assumptions, centered on essentialized and static understandings of concepts makes it difficult for researchers who locate themselves within a postpositivist paradigm to understand or value the contributions that poststructuralism makes in terms of critique and identification of the invisibility of mainstream discourses, practices, values, and beliefs. Poststructuralism makes people uncomfortable because researchers who use it embrace fluidity, dynamisms, and constantly changing language and meanings. Those located within a postpositivist paradigm, are comforted by certainty and often the uncertainty of poststructuralism can be threatening. This is challenging for researchers who use poststructuralism, within academic settings where postpositivism is the received view.

You can read more about Danielle Macdonald's research

Macdonald, D. (2019). *Exploring collaboration between midwives and nurses in Nova Scotia: A feminist poststructuralist case study.* [Doctoral Dissertation, University of Ottawa]. https://ruor. uottawa.ca/handle/10393/39112

Macdonald, D., & Etowa, J. (2021). Experiences of and visions for collaboration between midwives and nurses in Nova Scotia. *Women and Birth. 34*(5), e482–e492. https://doi.org/10.1016/j. wombi.2020.10.004

Macdonald, D. (2022). Relationships, roles, and person-centred practices – Collaborative birthing care in Nova Scotia. *International Practice Development Journal. 12*(1). https://doi.org/ 10.19043/ipdj.121.006

Dr. Sara Kirk PhD – Professor in the Faculty of Health Dalhousie University

I (Megan) met with Dr. Sara Kirk, a professor at Dalhousie University at a local coffee shop to talk about her experience using poststructural approaches and feminist thought in the study "Balancing the scales: obesity management." The following are excerpts from our conversation. To begin the conversation, I simply asked her to start anywhere she would like and tell me what she thought about using a poststructural approach in this research study.

Excerpt 1: Sara began by saying she had been "blown away" by the possibilities afforded by a poststructural approach because we had been able to apply it to the 'real world.' Sara was a quantitative researcher and had never used a poststructural approach. She recounted our first meeting when we discussed the possibility of using a poststructural approach in an obesity management study she was thinking about doing. She said she was intrigued by my confidence that poststructural thought would be the perfect qualitative methodology to use to explore obesity management from the perspective of people living with obesity, their health care providers, and policy makers. Sara said she was excited and although she knew nothing about poststructural thought, she was happy to learn along the way and trust the process. Before data collection was completed, we published a paper about how we were going to use a poststructural approach (Aston et al., 2011). Sara said that this was invaluable as it provided her with a road map to help her understand the processes of data collection and analysis. As we began collecting the data (participant stories about experiencing obesity and health care management through semi structured interviews) we started analyzing. Sara said that she could see how poststructural thought provided rigour and structure to analysis and she was able to understand the concepts because they had been brought to life through participant stories. It was at this point that her excitement about the possibilities of poststructural approaches combined with feminist thought really started. In fact, she used the phrase 'life changing' and emphasized her point by saying she was not embellishing by using this phrase. She genuinely believed that it was the insights from participants that made the research so powerful.

Excerpt 2: At the time of the study, nothing had been published that looked at obesity from a social justice lens that included negotiating power relations. We were on the cusp of looking at obesity in a new way, using the possibilities afforded by a poststructural approach. Indeed, when we conducted our study, it was quite revolutionary. Sara said she had definitely seen changes in the way language was shifting in the literature and in practice regarding obesity management and believed our research had influenced this. She said there was a 'shift in the dialogue.' She had received emails from people commending our work and stating that it had changed the way they understood obesity. Using insights afforded by poststructural thought had changed the way she interacted with health care systems. Overall, Sara said that we needed to talk more about power, as this helped us to really make a difference. Although she could only loosely understand the poststructural concept of power, she could absolutely see how power worked through the participants' experiences. She said it even gave her goosebumps! Sarás reflections highlight how using poststructural thought in research can really have an impact on how we live our lives and offer health care services.

Balancing the Scales: A performance to change health care practices

Sara Kirk, Sheri Price, and Megan Aston led the writing of a play based on findings from research. "Balancing the scales" was a 15 minute play/video used in workshops for health care professionals. Actors presented the play and the audience was given group tasks so they could critique and respond to the messages in the play and then rewrite the script to ensure there was a positive and empathetic relationship between health care professionals and clients. The first time the play went 'live' was with a group of health care professionals and some of the study participants who experienced obesity. The play was so powerful with many people being moved to tears and goosebumps! You can watch and read more about the obesity research. http://www.youtube.com/watch?v=LVX4_s5IP3g

If you want to read more about this study, check out the following articles.

Aston, M., Price, S., Kirk, S., Penney, T. (2011). More than meets the eye. Feminist poststructuralism as a lens towards understanding obesity. *Journal of Advanced Nursing* 68(5), 1187–1194. Doi:10.1111/j.1365-2648.2011.05866.x

Price, S., Sim, M., Kirk, S., Aston, M. & Awad, C (2017). An innovation arts-based approach to interprofessional education. *Health and International Practice* 3(2), eP1131 Available at: https://doi.org/10.7710/2159-1253.1131

Dr. Judy MacDonald PhD – Professor in the School of Social Work Dalhousie University

Dr. Judy MacDonald has combined poststructural, disability theory, and feminist thought through her academic career. In her published book, *Critical Clinical Social Work: Counterstorying for Social Justice*, she and a colleague share theoretical and practical discussions about the importance of attending to relations of power, the social construction of disabilities, and a focus on the (dis) Abled body as a site for contestation. The following are some excerpts from Megan's interview with Judy.

Excerpt 1: I asked Judy to meet with me to talk about her use of poststructural thought in her research. She began our conversation by saying "I don't usually use the term poststructuralism but I definitely use tenets of it." While her research was primarily about sharing personal experiences, with the purpose of 'counter-storying,' she also used ideas from Foucault in her practice, research, and teaching that challenged the status quo. As a Social Worker, Judy was passionate about applying theory to clinical practice with a focus on anti-oppression. Most of her work had focused on chronic pain and how to use story to oppose harmful biomedical practices. She told her own story of chronic pain in her PhD dissertation and had since supported numerous graduate students who wanted to conduct similar research telling their personal stories about chronic illness using a critical lens. Moving from feelings of losing control to gaining control had been a central focus of her work.

Excerpt 2: Judy explained that by using an autoethnography to deconstruct her chronic pain experience, and using concepts from poststructural thought, enabled her to 'shift the power balance' and 'regain power through her voice.' She also recognized that the dominant and sometimes oppressive biomedical discourse often prevented the voices of those experiencing chronic pain to be heard, and therefore, poststructural thought was one methodology that could help raise the voices of those who felt silenced and had not been recognized. She said that she did this by deconstructing medical discourses and truly respected and listened to the voices of participants.

Excerpt 3: Power, deconstruction, and feminist thought continued to help her understand personal experiences in the context of the health care system and society at large. Using concepts from both feminist and poststructural approaches had led her to a place of incorporating ideas from different disciplines. For example, the use of 'embodied experiences' from feminism helped her to critique the 'structural system.'

You can read more about Judy's work in her book

Brown, C. and MacDonald, J. (Eds) (2020). *Critical Clinical Social Work: Counterstorying for Social Justice*. Toronto ON: Canadian Scholars Press.

Dr. Matthew Numer PhD – Professor School of Health and Human Performance Dalhousie University

Excerpt 1: I met with Dr. Matthew Numer to talk about poststructural thought, how he was introduced to it, what it meant to him, and how he used it in his own work, research, teaching, and practice. Matthew said he was introduced to poststructural thought 'by accident' as a Masters' student to help him research modern masculinity. He explained that he used a poststructural approach to disrupt categories of 'man' in a way that was different from a feminist poststructural deconstruction of the category of 'woman.' While both were used to fight for equity, they called for different understandings of the social construction of male and female bodies and the use of power.

Excerpt 2: Matthew explained that when he taught his undergraduate course on 'Human Sexuality' it was imperative that he considered the diversity of the undergraduate students in his class and how people were socially positioned. He paid close attention to how he shared poststructural ideas particularly for those who were not familiar with the language of poststructural thought or activism. For example, he was mindful not to create blame towards groups of people as this could be polarizing and harmful. For example, poststructural thought enabled him to recognize the complexities of social construction and power such as white male privilege; therefore, he would never individualize his statements to cause a student to feel shamed or the cause of oppression. He chose his examples and the way he taught very carefully so as not to alienate students through the use of binary language.

Excerpt 3: Matthew then spoke about his use of poststructural approaches and discourse from a theoretical point of view. He said that one can only experience and react to discourses they know and gave the example of how gay men were challenging dominant heteronormative sex discourses with the use of hook-up apps such as Grindr. These dating apps for gay men created spaces to talk about and experience sex as a 'line of flight' or a 'line of escape' from the dominant heteronormative sex discourse. I was intrigued by this new use of language – language that was used to capture a poststructural concept as well as seek out different experiences and discourses that might have been 'hidden' or 'oppressed'; thereby offering new or modern possibilities of living in the world. Matthew also said he used poststructural thought as a way to push back against neoliberalism by bringing forth suppressed discourses. Understanding how using phones and hook-up apps constructed a new digital space was timely and a different way of understanding one's position in the world. The phone became the voice of the participants and the internet was a new way to freedom, yet at the same time created new constraints.

Excerpt 4: When explaining the statement 'gay can never be masculine' he said because gay men constantly compare their experiences to the dominant discourse of masculinity that is white, male gender, physically strong and emotionally detached, they will always be seen as 'other.' A discourse of gay men's sex needs to acknowledge many complexities.

If you would like more information on Dr. Numer's research and his use of poststructuralism you can read his publications

https://www.dal.ca/faculty/health/health-humanperformance/faculty-staff/our-faculty/health-promotion/matthew-numer.html

Dr. Lisa Goldberg PhD RN– Professor School of Nursing, Dalhousie University

The following are excerpts from an interview I conducted with Dr. Lisa Goldberg

Excerpt 1: Dr. Lisa Goldberg is a Caritas Scientist who conducts research in the area of Queer women's birthing experiences. During our interview, Lisa began by saying that all of the theories she used were 'living, breathing extensions of herself' and included feminism, caring science, phenomenology, queer theory, and some aspects of poststructuralism. Queer and feminist theories were important as her work was political, and while poststructural ideas were important, she didn't always use poststructural language, yet it was woven throughout. Concepts from Caring Science were also included such as 'relationships, trust, respect and love.'

Excerpt 2: Lisa said that she chose to use Queer theory because it addressed the invisibility of those who identified as queer. However, she also recognized that poststructural thought could be used to examine the invisible. She liked feminist and poststructural thought because they could be applied to understand the construction of socio-cultural aspects of everyday living through structures and language. She said similar to queer theory, feminism was a political theory, and could be used to examine 'harms.' In particular, it was important to focus on whose stories were invisible and being omitted, such as the critical stories of women, queer, and under-represented equity seeking groups. These omissions were harmful, and often constructed through language. Through a poststructural approach she deconstructed language in combination with feminist and queer theories. In particular, Lisa liked Foucault's phrase to 'question the dangerous' and that this poststructural point of view could be used to support an examination of 'harms' from the perspective of gender and sexual orientation.

Excerpt 3: Lisa believed it was important to examine experiences through a critique of language and discourse. For example, in her view heteronormative practices around birth had created homophobic conditions if one only used the language of 'birthing women' as this assumed that only certain people can birth and excludes others, including those who are trans and non-binary. She suggested that we needed to 'reimagine' what was possible in perinatal care, which in her opinion was beginning to happen. She said that it was important to search for alternative language and said that queer theory, feminist theory, and poststructural thought overlapped and could help to provide alternative language. Along with Foucault's phrase 'question the dangerous' Lisa also liked the feminist statement 'the personal is political' as it reminded her we are obligated to do something beyond theoretical analysis and provide advocacy and education that can change harmful practices; particularly with under-represented groups.

If you want to read more about Lisa Goldeberg's work some places to start are

Searle, J., Goldberg, L., Aston, M., & Burrow, S. (2017). Accessing new understandings of trauma-informed care with queer birthing women in rural Nova Scotia. *Journal of Clinical Nursing, 26*(21–22), 3576–3587. doi:10.1111/jocn.13727

Lisa Goldberg and Megan Aston co-produced a play 'What to expect when you're not expected' presenting findings from research that explored Queer women's birthing experiences in rural Nova Scotia.' You can find a description of the play in following book chapter.

Goldberg, L. & Aston, M. (2022). What to expect when you aren't expected: Bringing queer birthing lives from story to stage. In *Doing Performative Social Science: Creativity in Doing Research and Reaching Communities* (Ed Kip Jones). Routledge Taylor & Francis Group.

Evelyn Abudulai RN MScN Is Presently an Assistant Professor at Dalhousie University

Evelyn had recently graduated with her Masters of Science in Nursing when she wrote about her Masters thesis that used feminist and poststructural approaches. We have included excerpts from her writings.

Poststructural and Feminist Approaches Social Construction of Prenatal Obesity

Excerpt 1: The issue of body weight has been for decades, an enduring mainstay of contemporary scientific, social, and political discourse. The normalized widespread perception of a global obesity pandemic, continues to fuel a moral panic which has included putting spotlight on maternal bodies that are named obese.

Excerpt 2: Insight into how, why, for what purposes, and for whose benefit the discourse of obesity has been constructed or problematized has the potential to offer in general "a better understanding" of the multi-factorial, multi-contextual issue of obesity. With this in mind, a study inspired by a feminist poststructural approach was undertaken in a rural setting as part of my Masters study and demonstrated how the use of such an approach could illuminate and advance a better understanding of the lived experiences of two pregnant women who self-identified as obese.

Excerpt 3: Beyond facilitating an understanding of how they negotiated and positioned themselves within and between discourses, the study enabled insight into these women's often unwitting complicity in their own subjugation, and thereby their own inadvertent collusion in upholding and sustaining the constraining meaning they had come to associate with their experience of self, body, and pregnancy.

Excerpt 4: Thus, the FPS lens guided this study, not in the direction of unearthing some truth but to the more emancipatory project of interrogating how particular maternal obesity truths came to be, and what maintains those beliefs. The focus of inquiry or its point of reference is how particular knowledges come to inform and constitute particular experiences and practices. Therefore, using FPS enabled an analysis of these experiences in a manner that takes up feminist and power issues, thereby surfacing and exposing the relations of power inherent in what counts as truth, and what does not, about maternal obesity as well as what is marginalized and excluded and by what means.

Excerpt 5: The participants' experiences, captured in 3 themes (doing the best I can do, living in the public eye under public censure and not being listened to) brought to the surface the impact of every day taken-for-granted assumptions about maternal obesity based on the uncritical uptake of biomedical truths.

You can read Evelyn's Master's research thesis at Theses Canada
https://library-archives.canada.ca/eng/services/services-libraries/
theses/Pages/item.aspx?idNumber=1033225009

References

Aston, M., Price, S., Kirk, S., & Penney, T. (2011). More than meets the eye. Feminist poststructuralism as a lens towards understanding obesity. *Journal of Advanced Nursing, 68*(5), 1187–1194. https://doi.org/10.1111/j.1365-2648.2011.05866.x

Cheek, J. (1999). *Postmodern and Poststructural Approaches to Nursing Research.* Sage.

Collins, P.H. (1998). *Fighting Words: Black Women and the Search for Justice* (Vol. 7). University of Minnesota Press.

Collins, P.H., & Bilge, S. (2020). *Intersectionality.* John Wiley & Sons.

Crenshaw, K.W. (2017). *On Intersectionality: Essential Writings*. The New Press.

Foucault, M. (1975). *The Birth of the Clinic*. New York, Vintage Books.

Foucault, M. (1982). Afterword: The subject and power. In H. Dreyfus & P. Rabinow (Eds.), *Michel Foucault: Beyond Structuralism and Hermeneutics*. Chicago, University of Chicago Press, pp. 208–226.

Hooks, B. (1984). *Black Women Shaping Feminist Theory*. ProQuest Information and Learning.

Weedon, C. (1987). *Feminist Practice and Post Structuralist Theory*. London, Basil Blackwell.

Index